Research on Intelligent Manufacturing

Research on Intelligent Manufacturing (RIM) publishes the latest developments and applications of research in intelligent manufacturing—rapidly, informally and in high quality. It combines theory and practice to analyse related cases in fields including but not limited to:

Intelligent design theory and technologies
Intelligent manufacturing equipment and technologies
Intelligent sensing and control technologies
Intelligent manufacturing systems and services

This book series aims to address hot technological spots and solve challenging problems in the field of intelligent manufacturing. It brings together scientists and engineers working in all related branches from both East and West, under the support of national strategies like Industry 4.0 and Made in China 2025. With its wide coverage in all related branches, such as Industrial Internet of Things (IoT), Cloud Computing, 3D Printing and Virtual Reality Technology, we hope this book series can provide the researchers with a scientific platform to exchange and share the latest findings, ideas, and advances, and to chart the frontiers of intelligent manufacturing.

The series' scope includes monographs, professional books and graduate textbooks, edited volumes, and reference works intended to support education in related areas at the graduate and post-graduate levels.

If you are interested in publishing with the series, please contact Dr. Mengchu Huang, Senior Editor, Applied Sciences

Email: mengchu.huang@springer.com Tel: +86-21-2422 5094.

More information about this series at http://www.springer.com/series/15516

Wenfeng Wang · Hengjin Cai ·
Xiangyang Deng · Chenguang Lu ·
Limin Zhang

Interdisciplinary Evolution of the Machine Brain

Vision, Touch & Mind

華中科技大學出版社
Huazhong University of Science and Technology Press

Springer

Wenfeng Wang
Shanghai Institute of Technology
Shanghai, China

Xiangyang Deng
Naval Aeronautical University
Yantai, Shandong, China

Limin Zhang
Naval Aeronautical University
Yantai, Shandong, China

Hengjin Cai
School of Computer Science
Wuhan University
Wuhan, Hubei, China

Chenguang Lu
Institute of Intelligence
Liaoning Technical University
Liaoning, China

ISSN 2523-3386 ISSN 2523-3394 (electronic)
Research on Intelligent Manufacturing
ISBN 978-981-33-4243-9 ISBN 978-981-33-4244-6 (eBook)
https://doi.org/10.1007/978-981-33-4244-6

Jointly published with Huazhong University of Science and Technology Press
The print edition is not for sale in China (Mainland). Customers from China (Mainland) please order the print book from: Huazhong University of Science and Technology Press.

This Springer imprint is published by the registered company Springer Nature Singapore Pte Ltd.
The registered company address is: 152 Beach Road, #21-01/04 Gateway East, Singapore 189721, Singapore

To our family, for their unconditional love.
Wenfeng Wang
Hengjin Cai
Xiangyang Deng
Chenguang Lu
Limin Zhang

Preface

Artificial intelligence (AI) is changing the world. It was rapidly developed in the last decade and its importance for the human future has been widely recognized. However, brain-inspired intelligence is still a concept in debate, which can be largely attributed to the misunderstanding of the evolution law of the machine brain. Until now, scientific interpretation of machine intelligence is still not uniform. Obviously, scientists cannot simply equate brain-inspired intelligence with "automation". Is "machine intelligence" an alternative term of "brain-inspired intelligence"? or even, does it represent the whole story of AI? How to judge whether a machine (or robot) has brain-inspired intelligence? No widely recognized answers could be presented since the evolution law of machine brain remains poorly understood.

In the famous Turing test, a machine is defined to be intelligent if it could communicate with a human through telex devices, without being able to identify its machine identity. Based on this definition, intelligent machines should have brain-inspired intelligence, and therefore, machine intelligence is not the whole story of AI. The real machine intelligence represents the highest stage in the development and applications of AI. The natural way to approach this highest stage requires an interdisciplinary evolution of the machine brain—to develop learning systems through interdisciplinary researches.

Over the past decades, the development of AI has experienced three stages—machine computation, machine learning and machine understanding. Machine learning includes data mining. Environmental sensing helps to acquire the data. Both pattern analysis and scene understanding are significant parts of machine understanding. The development stage of AI also represents the current level of machine intelligence—cognitive computation. Scientists have taught machines how to collect and treat data and discover knowledge from the data. The evolution of machine brain will experience another two stages—machine meta-learning (learning to learn) and self-directed development (improving the capability of machine brain utilizing the learned knowledge).

This book aims to present a theoretical explanation on the above five stages to approach real machine intelligence, along with a preliminary understanding of the basic evolution law—interdisciplinary evolution of the machine brain. This means that any integration of machine learning with other subjects would potentially motivate the development of intelligence in the "machine brain". For the convenience of theoretical explanation, some published articles were included or re-organized in corresponding chapters of the book. We claim that Wei Zhao and Qi Xiong are co-authors of Chaps. 2 and 3, Ying Huang, Cong Huang and Ning Zhao are co-authors of Chaps. 5 and 6. Thanks to all the associated colleagues in Shanghai Institute of Technology, Chinese Academy of Sciences (Xinjiang Institute of Ecology and Geography), Wuhan University, Jimei University (Chengyi University College), Xi'an Jiaotong University, Naval Aeronautical University, Hunan University of Arts and Sciences and Liuzhou Railway Vocational Technical College for their significant contributions in the originally published journal articles. Thanks for the support of the National Natural Science Foundation of China (41571299) and the Shanghai High-level Base-building Project for Industrial Technology Innovation (1021GN204005-A06).

Shanghai, China — Wenfeng Wang
Wuhan, China — Hengjin Cai
Yantai, China — Xiangyang Deng
Vancover, Canada — Chenguang Lu
Yantai, China — Limin Zhang
June 2020

Contents

Chapter 1
Introduction of Cognition Science

Abstract To understand the basic evolution law of the machine brain, we need first understand machine cognition, which majorly depends on machine vision, machine touch and etc. Artificial intelligence (AI) has been rapidly developed in the latest decade and its importance to machine cognition has been widely recognized. But machine minds is still a dream about the unknowable future. This chapter presents a perspective characterization of machine cognition in the future, taking the future of medical rescues as a practical example. A conjecture on key points of theoretical framework of machine cognition processes in the considered scenes—the concept of "ambulanceye" is advanced. Ambulanceye incorporates vision, touch and minds of the future intelligent ambulances with geospatial artificial intelligence (geoAI). The role of geoAI is hypothesized as a significant contributor to driving security through real-time danger caution. Such danger caution is based on the advanced driver assistant systems (ADAS) for detecting and tracking "eigenobjects", which are defined as potential dangerous objects hidden in spatial data. Simulated performances demonstrate the significance of ADAS for ambulances to overcome ocular restriction resulted from weathering extremes and other accidents that make sights blurred. Moreover, there are still considerable uncertainties in real-time recognition and the trace characterization of eigenobjects. GeoADAS (Improved ADAS with geoAI) is hence required for more effective objects recognition and tracking. Perspective applications of geoADAS are preliminarily discussed and at the end of this chapter, the outline of this book is also presented.

1.1 Background

Artificial intelligence (AI) is changing the world. It was rapidly developed in the latest decade and its importance for the human future have been widely recognized. However, brain-inspired intelligence is still a concept in debate, which can be largely attributed to misunderstanding of the basic evolution law of the machine brain. To understand the basic evolution law of the machine brain, we need first understand machine cognition, which majorly depends on machine vision, machine touch

W. Wang et al., *Interdisciplinary Evolution of the Machine Brain*,
Research on Intelligent Manufacturing,
https://doi.org/10.1007/978-981-33-4244-6_1

and etc. On one hand, AI has been rapidly developed in the latest decade and its importance to machine cognition has been widely recognized, on the other hand, machine minds is still a dream about the unknowable future. Here we will take a brief journey into the unknowable future, organize a perspective characterization of machine vision, touch and minds, taking the future of medical rescues as a practical example.

Intelligent monitoring system has been installed over the streets in the major cities and studies on the algorithms for the intelligent video recognition have attracted much attention [1–5]. Analysis of video bigdata has become one key direction associated with cities security and has been rapidly developed in the past decade, especially after its combination with geographic information, which further produced some new technologies—including real-time behavior recognition, information processing for 3D-reconstruction, exact acquirement of geographic data [6–10], geospatial artificial intelligence (geoAI) and etc.

This encourage us to offer a new blueprint for the future medical rescues to save driving time of ambulances in high speed and meanwhile, to ensure the security [11]. Considering rapid development of video recognition technology and, more and more attention are attracted to its applications in the advanced driver assistant systems (ADAS) in the latest decade [12–15]. Intelligent monitoring system with efficient facilities have been put into use in smart hospitals towards efficient medical rescues [16–20]. It is feasible to expand applications of AI to ambulances for medical rescues to strengthen the security of doctors, nurses, patients, drivers and the safety of other people and mobiles. Currently, the significance of ADAS to ambulances employed for medical rescues still has not attracted enough attention. Improving the drivers experiences by ADAS as a first approach is worthy of attempt [21–26]. In the future, incorporating ADAS with geoAI should be a next research priority towards a double-monitored driving security along the way to the scene of medical rescues. Alternatively, incorporated ADAS with geoAI (geoADAS) would be the next generation of ADAS for medical rescues.

Objectives of this chapter are: (1) to advance the concept of "ambulanceye" as a conjecture on the future medical rescues—assuming that both ADAS and geoAI can be equipped in ambulances and make contributions to driving security through timely danger recognition and caution, (2) to present some key points of geoADAS cognition processes and suggest geoADAS as a future research priority for the ADAS researchers, and (3) to organize a perspective characterization of machine vision, touch and minds from (1) and (2), along with the outline of this book.

1.2 Preliminaries and Basic Concepts

Geospatial videos were collected by the DVR in a driving car and this car is a hypothetical ambulance. Performances of ADAS in ambulances for medical rescues are simulated as understanding the scenes in these videos and finding potential dangers.

Danger recognition and caution are based on detecting and tracking of "eigenobjects", where "eigenobjects" are defined as the potential dangerous objects in these videos. A set of algorithms—optical flow method, compressive sensing tracking, deep learning [27–39] and geoAI are utilized in video recognition for validating feasibility of dangers recognition and caution.

1.2.1 *Region to Collect Geospatial Data*

In order to make it more challenging and convincing, geospatial videos are collected from the Hetian Region during sampling days with poor atmospheric visibility. The major geospatial information of Hetian Region is as follows. It is located in the southernmost part of Xinjiang, China. Xinjiang locates in the northwestern border of China, which is the largest province-level region in China division, with an arid climate, spanning over 1.6 million km^2 from the east border to the northwestern border of the Silk Road. Geographic location of Xinjiang in China and geographic location the Hetian Region in Xinjiang are shown in Fig. 1.1.

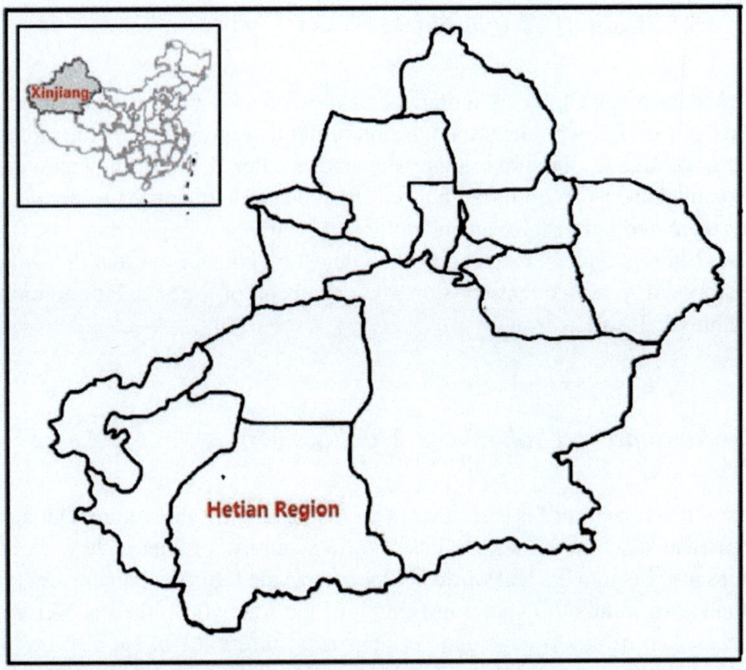

Fig. 1.1 Geographic locations of Xinjiang and the Hetian Region in China

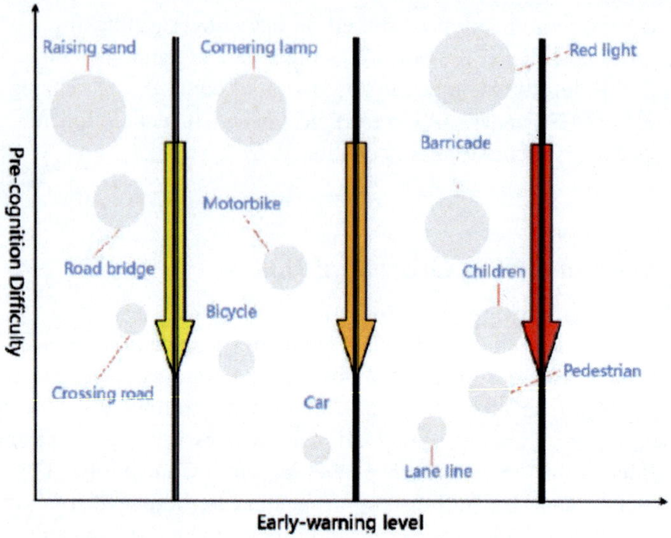

Fig. 1.2 A first sketch of eigenobjects implying potential dangers

1.2.2 Definition of "Eigenobjects"

A first sketch of eigenobjects along the way to the scene of medical rescues is presented in Fig. 1.2, with the pre-cognition difficulty represented by the circles size and early-warning level represented by the arrows color. As well-known, in medical rescues ambulances can go through a red light. This might caused traffic accidents and therefore, red light is also an eigenobject for early-warning of dangers. Overall, this first sketch includes 12 eigenobjects. It must be pointed out that this is not the whole story—any subsequent revision and expansion of the sketch of eigenobjects are certainly valuable.

1.2.3 Attempts to Understand Machine Minds

Until now, machine minds is still a dream about the unknowable future. Our attempts to understand machine minds are based on a scientific problem—how to expand machines insights into the real world? Hypothesize the future intelligent ambulances have their own minds to expand insights into the real world, then geoAI will not only be used to fill knowledge gaps in geospatial information, but also be used to present more efficient and effective solutions for specific problems. For example, geoAI can be utilized to predict which areas of the city will be faced with extreme traffic congestion, what kind of solutions to be taken, how to select the route of ambulances and so on.

Such prediction certainly makes sense for drivers to know how serious the traffic problems could be and where these problems might happen. In the past decades, geospatial science has made great progress and already satisfies the requirement for developing geoAI for unmanned driving and even, next generation of automation. Simulated experiments about the dangers detection and caution will be carried out in Sect. 1.3.

Machine minds are theoretically feasible to be realized. The human mind depends on the brain, where the brain is made of matter. The future machine is an artificial human and the machine brain can be understand as an artificial brain. They will have ability to make decisions on how to expand their own insights for recognition of eigenobjects in the future medical rescues.

1.3 Perspective Simulation and Characterization

As well-known, any driving car has a blind area and ambulances are not an exception. ADAS will help drivers to avoid accidents when they are driving ambulances to the scene of medical rescues. We define ADAS for the future ambulances as an "ambulanceye" and conjecture that ambulanceyes have not a blind area. Nobody can deny the necessity to utilize ambulanceye in the future unless they could present another efficient way to avoid unfortunate accidents (although occasionally) along the way to the scene of medical rescues. The basic idea in practical applications of ambulanceye can be very simple—dangers recognition and caution.

1.3.1 Eigenobjects Detecting and Tracking

Examples of eigenobjects detecting and tracking indicate that ADAS can overcome the ocular restriction when facing weathering extremes (Fig. 1.3).

Ambulanceye as an emerging technology for monitoring ambulance driving safety in the future medical rescues can also be taken as a further combination of intelligent video recognition and real-time objects tracking with geoAI. Geospatial data and equipped geoADAS in ambulances are advantageous to timely detection and caution of potential dangers within the blind areas.

To further explain how geoAI help to find dangers and make a precaution for drivers, one most simple example—the performance of ambulanceye in applications to detect and warn lane-departure are validated for timely danger caution. The drivers learned well about the geographical information and so he is very careful in lane-departure detection and tracking. This is simulated performances of geoAI incorporated in the next generation of ADAS for dangers recognition and caution in driving ambulances. As seen in Fig. 1.4, geoAI helps ADAS to break restriction of accidents that causes sights blurred.

Fig. 1.3 Examples of eigenobjects detecting and tracking

It is also necessary to point out that videos collected by vehicle-tachograph need a 'debouncing' treatment, which improved the simulated performances.

1.3.2 Necessity to Introduce GeoADAS

Integrating the widely-recognized insights, a full screen of the blind areas of drivers in ambulances can be characterized in Fig. 1.5.

Furthermore, importance of geoADAS shall be largely recognized since the treatments of additional accidents cost much time. It is easy to see that the vast majority of the blind areas are in the lack of realization by the ambulance drivers, because that the medical rescues are always emergent and their time maybe insufficient. If spending too much time in artificially finding dangers, they might cannot arrive the scene of medical rescues in time, which implies that a great loss of people's lives and property would be produced. geoADAS is suggested to be employed in ambulances to reduce additional unfortunate accidents. Consequently, ambulanceye is very important for improving the efficiency of intelligent medical rescues in the

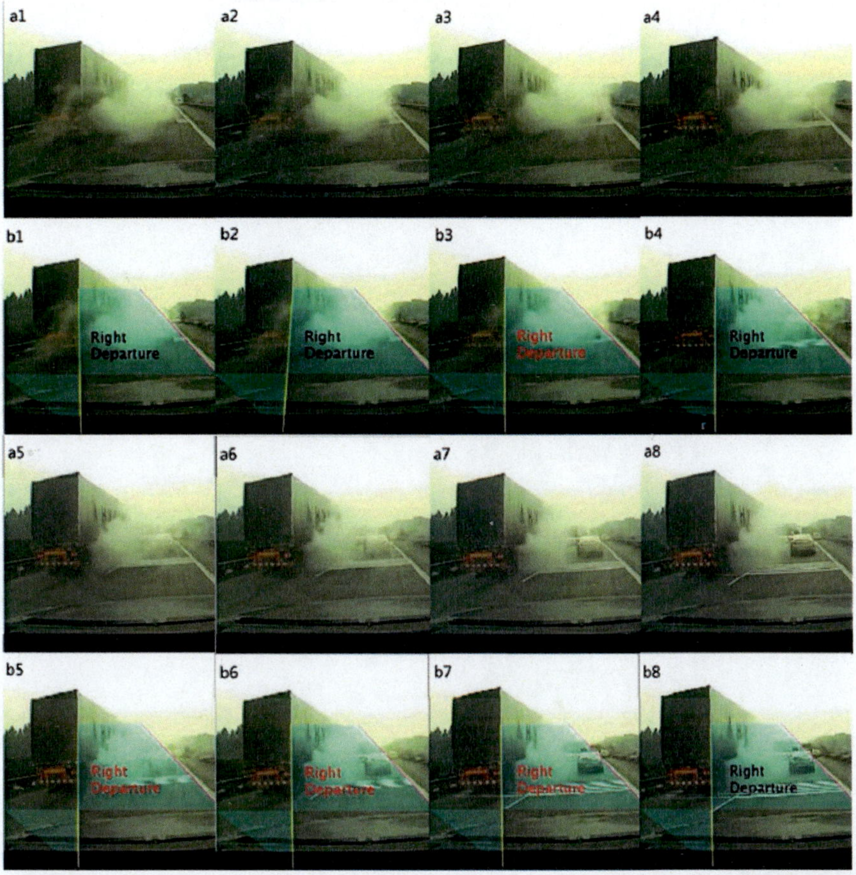

Fig. 1.4 Lane departure detecting for danger caution

future. In short, ambulanceye requires a geoADAS system for real-time recognition of eigenobjects.

1.3.3 A Perspective Characterization of Ambulanceye

A first perspective characterization on the machine cognition processes of an ambulanceye is shown in Fig. 1.6. A part of the doctors and nurses in the future intelligent ambulances will be artificial human. They have an artificial brain. They will have machine vision, touch and minds. They will be able to make a great help to human in the future medical rescues.

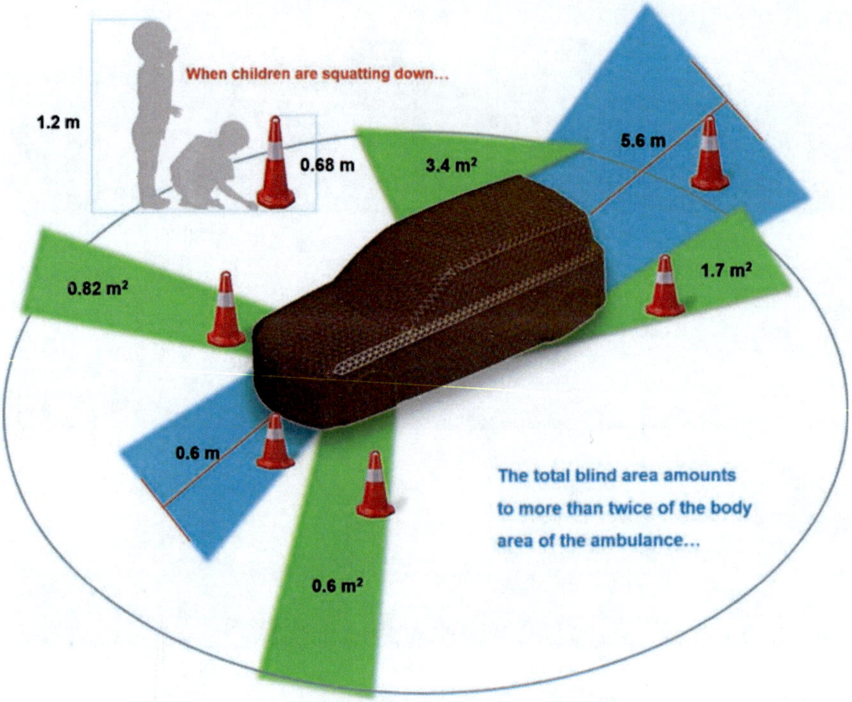

Fig. 1.5 Description of blind areas of drivers in ambulances

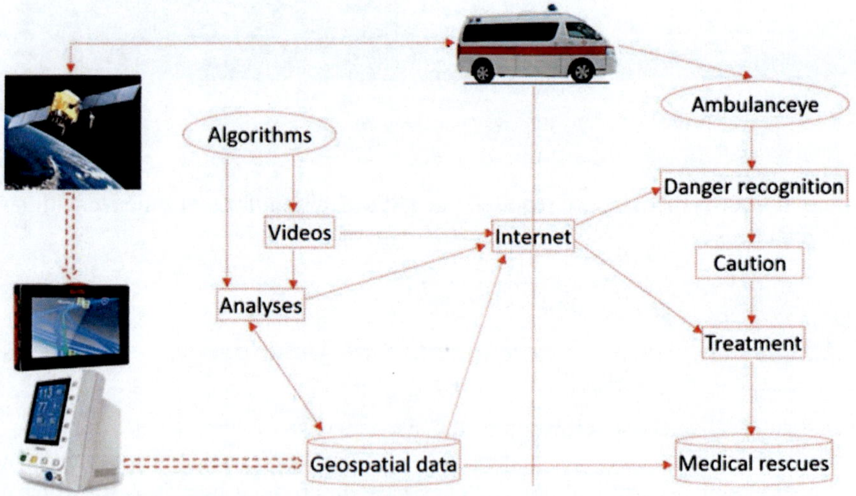

Fig. 1.6 A perspective characterization: cognition processes of an ambulanceye

1.4 Theoretical Framework of Machine Cognition

Geographical environment is essential significant for machine cognition in the future. Environment is not only the basis for human survival and science development, but also the material basis for machine intelligence. Therefore, the role of geoAI in cognition science is of great significance, which is mainly reflected in the evolution of machine brain. It is inevitable that machines and their activities in this environment have a process of production, development and extinction. All machine cognition processes must conform to the internal laws of the corresponding geospatial environment. We wish the machine brain could be equipped with geoAI and in the future, help human to improve the environments according to human needs and subsequently, delay or stop some transformation of materials and some transmission of energy.

Cognition science with geoAI will also help machine to understand why their activities have different results in different regions and in different periods. As seen in this chapter, ambulanceye as an emerging technology can detect and track eigenobjects and help to break the restriction of geographical environment on the efficiency of medical rescues. It will facilitate the visualization of potential dangers even when facing special weathering accidents and other unexpected accidents that causes sights blurred. GeoAI will not only link doctors, nurses, drivers and the archival data of local information for danger recognition and caution along the way to the scene of medical rescues, but also expands applications of ADAS.

Therefore geoADAS will enhance the security experience of the doctors, nurses, patients and drivers in the future medical rescues and in turn, improve work efficiency. For theoretical illustration, a simple yet efficient model of working time in one medical rescue can be formulated as follows

$$t = b - c * x\%$$

where t is the practically working time in a medical rescue, b is the expected time, c is the technically condensed time by geoADAS and x% represents the efficiency and accuracy of geoAI in danger recognition and caution.

Obviously, efficiency of future medical rescues is largely determined by the coefficients c and x%. GeoAI indicates more efficient algorithms to reduce unfortunate accidents when ambulances are going to the scene of accidents. Furthermore, geoADAS can also help to improve efficiency of the future medical rescues and build reputation of smart hospitals—to help the people understand how the functions of ambulances are carried out wholly under the premise of ensuring safety—people can evaluate them by quantitative indices.

Cognition science involving geoAI for intelligent ambulances in the future can be further developed and make help in medical rescues. A next research priority is to find an optimal method for robust and real-time analyzing and mining the history and personnel movement trace. A first framework of the future medical rescues is shown in Fig. 1.7.

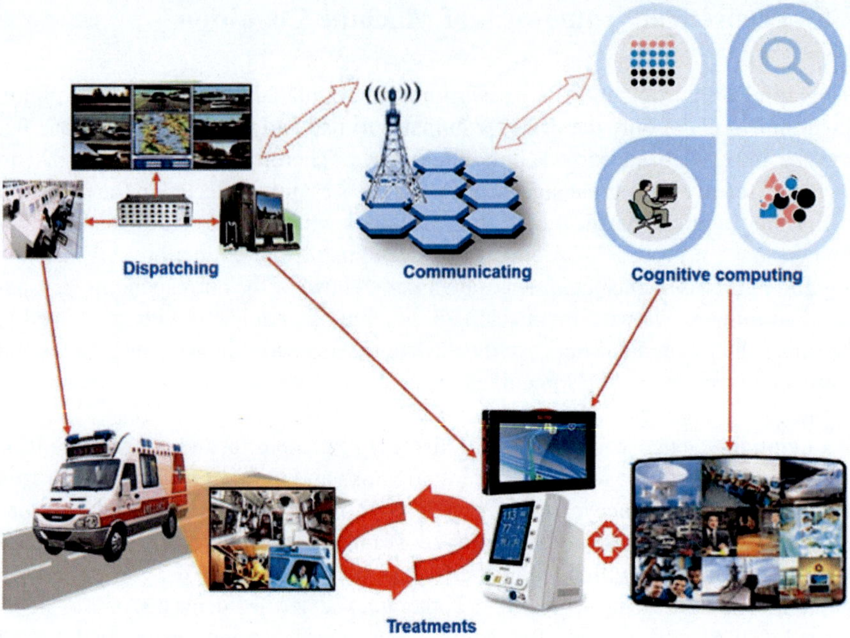

Fig. 1.7 Theoretical framework of geoADAS cognition: ambulanceye as an example

Statements in Sects. 1.1–1.4 were explicitly re-organized from [40]. This is not a dream. Current machines from different countries have quite different degrees of cognition to the geographical environment and different ways of utilization of geospatial information. The evolution of machine brain is bound to be interdisciplinary, colorful and diverse.

1.5 Outline of the Book

Over the past decades, the development of AI have experienced 3 stages—machine computation, machine learning and machine understanding. Machine leaning includes data mining. Environmental sensing helps to acquire the data. Both pattern analysis and scene understanding are significant parts of machine understanding. The development stage of AI also represent the current level of machine intelligence—cognitive computation. Scientists have taught machine how to collect and treat data and discover knowledge from the data. Evolution of machine brain will experience another two stages—machine meta-learning (learning to learn) and self-directed development (improving the capability of machine brain utilizing the learned knowledge). This book aims to present a theoretical explanation on the above five stages to approach real machine intelligence, along with a preliminary understanding of the basic evolution law—interdisciplinary evolution of the machine brain. Organization of the remaining chapters of this book is as follows.

Chapter 2 explains how to construct a machine learning system and summarize the current level of machine intelligence as cognitive computation. One-dimensional deep neural network is utilized for illustration and detection of recorded epileptic seizure activity in Electroencephalogram (EEG) segments is given as a practical application. Manual recognition is a time-consuming and laborious process so that it places a heavy burden on neurologists, and hence, the automatic identification of epilepsy has become an important issue. Such intelligent detection is crucial for the classification of seizures. Traditional EEG recognition models depend on artificial experience and are of weak generalization ability. To break these limitations, we propose a novel one-dimensional deep neural network for robust detection of seizures, which consists of three convolutional blocks and three fully connected layers. Thereinto, each convolutional block consists of five layers: convolutional layer, Batch normalization layer, non-linear activation layer, dropout layer, and max-pooling layer. Machine cognition performance is evaluated on the University of Bonn dataset, which achieves an accuracy of 97.6–99.5% in the two-class classification problem, 93.7–98.1% in the three-class EEG classification problem, and 93.6% in classifying the complicated five-class problem.

Chapter 3 explains the processes of machine cognition for a better understanding of environmental changes at the current level of machine intelligence and conjecture how evolution of the machine brain would change the future way of knowledge discovery (data mining) in environments sensing. In order to strengthen the continuity of Chap. 2, we present a parallel framework based on MPI for large dataset to extract power spectrum features of EEG signals so as to improve the speed of brain signal processing. At present, the Welch method has been wildly used to estimate the power spectrum. However, the traditional Welch method takes a lot of time especially for large dataset. In view of this, we added the MPI into the traditional Welch method and developed it into a reusable Master–Slave parallel framework. As long as the EEG data of any format is converted into the text file of a specified format, the power spectrum features can be extracted quickly by this parallel framework. In the proposed parallel framework, the EEG signals recorded by a channel are divided into N overlapping data segments. Then the PSD of N segments are computed by some nodes in parallel. The results are collected and summarized by the master node. The final PSD results of each channel are saved in the text file, which can be read and analyzed by the Microsoft Excel. This framework can be implemented not only on the clusters but also on the desktop computer. In the experiment, we deploy this framework on a desktop computer with a 4-core Intel CPU. It took only a few minutes to extract the power spectrum features from 2.85 GB EEG dataset, seven times faster than using the Python. This framework makes it easy for users, who do not have any parallel programming experience in constructing the parallel algorithms to extract EEG power spectrum. At the end of this chapter, reinforcement learning for solving a blind separation problem is carried out to highlight the necessity to change the way of data mining in environments sensing.

Chapter 4 explains the pattern of machine understanding, utilizing medical test as a practical example. The explanation is based on the semantic information theory. After long arguments between positivism and falsificationism, the verification of universal hypotheses was replaced with the confirmation of uncertain major premises. Unfortunately, Hemple proposed the Raven Paradox. Then, Carnap used the increment of logical probability as the confirmation measure. So far, many confirmation measures have been proposed. Measure F proposed by Kemeny and Oppenheim among them possesses symmetries and asymmetries proposed by Elles and Fitelson, monotonicity proposed by Greco et al., and normalizing property suggested by many researchers. Based on the semantic information theory, a measure b^* similar to$F$ is derived from the medical test. Like the likelihood ratio, measures b^* and F can only indicate the quality of channels or the testing means instead of the quality of probability predictions. Furthermore, it is still not easy to use b^*, F, or another measure to clarify the Raven Paradox. For this reason, measure c^* similar to the correct rate is derived. Measure c^* supports the Nicod Criterion and undermines the Equivalence Condition, and hence, can be used to eliminate the Raven Paradox. An example indicates that measures F and b^* are helpful for diagnosing the infection of Novel Coronavirus, whereas most popular confirmation measures are not. Another example reveals that all popular confirmation measures cannot be used to explain that a black raven can confirm "Ravens are black" more strongly than a piece of chalk. Measures F, b^*, and c^* indicate that the existence of fewer counterexamples is more important than more positive examples' existence, and hence, are compatible with Popper's falsification thought.

Chapter 5 presents a theoretical framework on the evolution stages of the machine brain and cognitive computation and systems for machine computation, learning and understanding. We divide AI subject into 2 branches—pure AI and applied AI (defined as an integration of AI with another subject: geoAI as an example). To stretch the continuation of this chapter, we first analyze how to predict dangers in unmanned driving with geoAI and introduce the robot path planning (RPP) problem. Subsequently, an ant colony optimization (ACO) algorithm for solving the RPP problem are interpreted to understand cognitive computation and systems for machine computation, learning and understanding. A practical example of RPP problem—the traveling salesman problem (TSP) is further introduced. Integrating ACO with the iteration-best pheromone update rule, the ACO algorithm is improved and an adaptive mechanism are presented to treat instability. Experiments show that the new ACO algorithm has a good performance and robustness. Stability of the cognitive system and its robustness in cognitive computation for solving TSP are further validated. The vision-brain hypothesis, which has been proposed in the book "Brain-inspired intelligence and Visual Perception", is developed and hence extended as the vision-minds brain hypothesis. At the end of this chapter, as a first theoretical utilization of the vision-minds brain hypothesis, we explain how artificial improvements of the algorithms in applied AI can contribute to evolution of the machine brain.

Finally, Chap. 6 finally presents a characterization of interdisciplinary evolution of the machine brain. Perspective schemes for rebuilding a real vision brain in the future are analyzed, along with the major principles to construct the machine brain, are presented, which include memory, thinking, imagination, feeling, speaking and other aspects associated with machine vision, machine touch and machine minds. This explicitly developed the theoretical framework of brain-inspired intelligence from the vision brain hypothesis, the vision-minds hypothesis and the skin brain hypothesis. Based on Chaps. 2–5, development of machine intelligence during the past decades have experienced three stages—machine computation, learning and understanding. Machine leaning includes data mining. Environmental sensing helps to acquire the data. Both pattern analysis and scene understanding are significant parts of machine understanding. Scientists have taught machine how to collect and treat data and discover knowledge from the data. Evolution of machine brain will experience another two stages—machine meta-learning (learning to learn) and self-directed development (improving the capability of machine brain utilizing the learned knowledge). There are still great challenges in realization of the dream.

References

1. Kovar L. and Gleicher M. Automated extraction and parameterization of motions in large data sets [J]. Acm Transactions on Graphics, 2004, 23(3): 559–568.
2. Philip Chen C. L. and Zhang C. Y. Data-intensive applications, challenges, techniques and technologies: A survey on Big Data [J]. Information Sciences, 2014, 275(11): 314–347.
3. Marx V. The big challenges of big data [J]. Nature, 2013, 498(7453): 255–260.
4. Richtárik P. and Takáč M. Parallel coordinate descent methods for big data optimization [J]. Mathematical Programming, 2016, 156(1): 1–52.
5. Vazhkudai S. and Schopf J. M. Using Regression Techniques to Predict Large Data Transfers [J]. International Journal of High Performance Computing Applications, 2003, 17(3): 249–268.
6. Waldrop M. Big Data: Wikiomics [J]. Nature, 2008, 455(7209): 22–25.
7. Kim G. H., Trimi S. and Chung J. H. Big-Data Applications in the Government Sector [J]. Communications of the Acm, 2014, 57(3): 78–85.
8. Mervis J. Agencies Rally to Tackle Big Data [J]. Science, 2012, 336(6077): 22–22.
9. Wigan M. R. and Clarke R. Big Data's Big Unintended Consequences [J]. Computer, 2013, 46(6): 46–53.
10. Talia D. Clouds for Scalable Big Data Analytics [J]. Computer, 2013, 46(5): 98–101.
11. Westgate B. S., Woodard D. B., Matteson D. S. and Henderson S. G. Large-network travel time distribution estimation for ambulances [J]. European Journal of Operational Research, 2016, 252(1): 322–333.
12. Shin D. M., Yoon B. G., Han Y. T. Analysis of Ambulance Traffic Accident During Driving [J]. 2016, 30(1):130–137.
13. Shin D. M., Kim S. Y., Han Y. T. A Study on the Comparative Analysis of Fire-Fighting Ambulances about the Aspects of Safety and Efficiency using the Question Investigation [J]. 2015, 29(2):44–53.
14. Ambrose J. Emergency response driving education within UK ambulance services [J]. Journal of Paramedic Practice, 2013, 5(6):351–353.
15. Sundström A, Albertsson P. Self- and peer-assessments of ambulance drivers' driving performance [J]. Iatss Research, 2012, 36(1):40–47.
16. Raaber N., Duvald I., Riddervold I., Christensen E. F. and Kirkegaard H. Geographic information system data from ambulances applied in the emergency department: effects on patient reception [J]. Scandinavian Journal of Trauma, Resuscitation and Emergency Medicine, 2016, 24(1): 1–9.

17. Fu Q., Li B., Yang L., Wu Z. and Zhang X. Ecosystem services evaluation and its spatial characteristics in central asia's arid regions: a case study in altay prefecture, china. Sustainability, 2015, 7(7): 8335–8353.
18. Xie Z. and Liu G. Blood perfusion construction for infrared face recognition based on bio-heat transfer [J]. Bio-medical materials and engineering, 2014, 24(6): 2733–2742.
19. Jin L., Niu Q., Jiang Y., Xian H., Qin Y. and Xu M. Driver sleepiness detection system based on eye movements variables. Advances in Mechanical Engineering, 2013, 2013(5): 1–7.
20. Wang T., Dong J., Sun X., Zhang S. and Wang S. Automatic recognition of facial movement for paralyzed face. Bio-medical materials and engineering, 2014, 24(6): 2751–2760.
21. Vithya G., Sundaram B. V. Inpatient Critical Stage Monitoring in Smart Hospitals by Contextual Fuzzy based QoS Routing for WBMS Network Nurse Call System [J]. Wireless Personal Communications, 2016:1–16.
22. Nandyala C. S., Kim H K. From Cloud to Fog and IoT-Based Real-Time U-Healthcare Monitoring for Smart Homes and Hospitals [J]. International Journal of Smart Home, 2016, 10(2):187–196.
23. Chen X., Wang L., Ding J., et al. Patient Flow Scheduling and Capacity Planning in a Smart Hospital Environment [J]. IEEE Access, 2016, 4:135–148.
24. Al-Refaie A., Chen T., Judeh M. Optimal operating room scheduling for normal and unexpected events in a smart hospital [J]. Operational Research, 2016:1–24.
25. Vecchia G. D., Gallo L., Esposito M., et al. An infrastructure for smart hospitals [J]. Multimedia Tools and Applications, 2012, 59(1):341–362.
26. Yao W., Chu C. H., Li Z. Leveraging complex event processing for smart hospitals using RFID [J]. Journal of Network & Computer Applications, 2011, 34(3):799–810.
27. Fang Y. L., Zhang A., Wang H., Li H., Zhang Z.W., Chen S.X., Luan L.Y. Health risk assessment of trace elements in Chinese raisins produced in Xinjiang province [J]. Food Control, 2010, 21(5): 732–739.
28. Jing L. Incremental Learning for Robust Visual Tracking [J]. International Journal of Computer Vision, 2008, 77(1-3):125–141.
29. Dewan M. A. A., Granger E., Marcialis G. L., et al. Adaptive appearance model tracking for still-to-video face recognition [J]. Pattern Recognition, 2016, 49(C):129–151.
30. Babenko B., Yang M. H., Belongie S. Robust Object Tracking with Online Multiple Instance Learning [J]. IEEE Transactions on Pattern Analysis & Machine Intelligence, 2011, 33(8): 1619–1632.
31. Wu Y., Jia N., Sun J. Real-time multi-scale tracking based on compressive sensing [J]. Visual Computer International Journal of Computer Graphics, 2015, 31(4): 471–484.
32. Mei X., Ling H. Robust Visual Tracking and Vehicle Classification via Sparse Representation [J]. IEEE Transactions on Software Engineering, 2011, 33(11): 2259–2272.
33. Yamins D. L. K. and Dicarlo J. J. Using goal-driven deep learning models to understand sensory cortex [J]. Nature Neuroscience, 2016, 19(3): 356–365.
34. Chen L., Qu H., Zhao J. and Principe J. C. Efficient and robust deep learning with Correntropy-induced loss function [J]. Neural Computing and Applications, 2016, 27(4):1019–1031.
35. Ghesu F. C., Krubasik E., Georgescu B. and Singh V. Marginal Space Deep Learning: Efficient Architecture for Volumetric Image Parsing [J]. IEEE Transactions on Medical Imaging, 2016, 35(5):1217–1228.
36. Erfani S. M., Rajasegarar S., Karunasekera S. and Leckie C. High-dimensional and large-scale anomaly detection using a linear one-class SVM with deep learning [J]. Pattern Recognition, 2016, 58: 121–134.
37. Greenspan H., Ginneken B. V. and Summers R. M. Guest Editorial Deep Learning in Medical Imaging: Overview and Future Promise of an Exciting New Technique [J]. IEEE Transactions on Medical Imaging, 2016, 35(5): 1153–1159.
38. Wang Y., Luo Z. and Jodoin P. M. Interactive Deep Learning Method for Segmenting Moving Objects [J]. Pattern Recognition Letters, 2016, https://doi.org/10.1016/j.patrec.2016.09.014.
39. Ngo T. A., Lu Z. and Carneiro G. Combining Deep Learning and Level Set for the Automated Segmentation of the Left Ventricle of the Heart from Cardiac Cine Magnetic Resonance [J]. Medical Image Analysis, 2016, 35: 159–171.

40. Wang W. F., Chen X., Zhou H. Y., et al. Ambulanceye – The Future of Medical Rescues[C]. The 3rd International Conference on Cognitive Systems and Information Processing, Springer Singapore, 2017, 606–615.

Chapter 2
Cognitive Computation and Systems

Abstract This chapter aims to explain how to construct a machine learning system and summarize the current level of machine intelligence as cognitive computation. One-dimensional deep neural network is utilized for illustration and detection of recorded epileptic seizure activity in Electroencephalogram (EEG) segments is given as a practical application. Manual recognition is a time-consuming and laborious process so that it places a heavy burden on neurologists, and hence, the automatic identification of epilepsy has become an important issue. Such intelligent detection is crucial for the classification of seizures. Traditional EEG recognition models depend on artificial experience and are of weak generalization ability. To break these limitations, we propose a novel one-dimensional deep neural network for robust detection of seizures, which consists of three convolutional blocks and three fully connected layers. Thereinto, each convolutional block consists of five layers: convolutional layer, Batch normalization layer, non-linear activation layer, dropout layer, and max-pooling layer. Machine cognition performance is evaluated on the University of Bonn dataset, which achieves an accuracy of 97.6–99.5% in the two-class classification problem, 93.7–98.1% in the three-class EEG classification problem, and 93.6% in classifying the complicated five-class problem.

2.1 Background

Cognitive computation represents the current development stage of AI and also is the current level of machine learning. One-dimensional deep neural network is a widely applied cognitive system. It is always utilized to describe the cognitive computation mechanisms of deep learning systems. So we will utilize it to explain how to construct a cognitive system in machine learning and cognitive computation. Intelligent detection of recorded epileptic seizure activity in Electroencephalogram (EEG) segments is employed as a practical application. Electroencephalogram (EEG) is a non-invasiveness, effective technique used in clinical studies to decode the electrical activity of the brain. EEG is one of the critical technologies to identify an abnormality of the brain, such as detecting epileptic seizures. Seizures are transient neurological

W. Wang et al., *Interdisciplinary Evolution of the Machine Brain*,
Research on Intelligent Manufacturing,
https://doi.org/10.1007/978-981-33-4244-6_2

dysfunctions caused by abnormal brain neurons and excessive super synchronized discharges. The visual inspection of EEG for seizure detection by expert neurologists is a time-consuming and laborious process, and the diagnosis may not be accurate because of the massive amounts of EEG data and the discrepant clinical judgment standards of different neurologists [1, 2]. Therefore, scientific research on EEG-based automatic detection of epilepsy has attracted much attention.

Recently, numerous algorithms have been proposed in the literature to automatic detection of an epileptic seizure by using different features of the EEG signals. These methods can be roughly classified into two categories: conventional methods and deep learning (DL) based methods. Thereinto, most of the traditional methods use hand-engineered techniques for feature extraction from EEG signals, then conjunct with classifiers to recognize. The Bonn University EEG database is widely used, which is publicly available and labeled as A, B, C, D, and E. Details of the dataset are described in a later section. There is much-published work using the Bonn dataset for epilepsy detection. They concern three main classification problems: the two-class seizure detection problem focus on the classification between non-seizures and seizures; the three-class epileptic classification problem concern on the grouping of three different EEG categories (normal, inter-ictal and ictal); the five-class recognition problem concern on the classification of five distinct types (A, B, C, D, and E).

In 2009, Ocak [3] proposed a scheme for detecting epileptic seizures based on approximate entropy and discrete wavelet transform (DWT) of EEG signals. This framework obtained an accuracy of 96% for two-class EEG classification. Moreover, Tzallas et al. [4] demonstrated the suitability of the time–frequency analysis (TFA) to classify EEG segments for epileptic seizures. The authors employed the artificial neural network (ANN) as the classifier and achieved an accuracy of 100% for the two-class and three-class classification, and 89% for the five-class rating. In 2010, Subasi et al. [5] employed principal components analysis, independent components analysis, and linear discriminant analysis to reduce the dimension of EEG signals, and extract statistical features from DWT, and then used support vector machine (SVM) for classification. This model yielded a seizure detection accuracy of 100% for two-class classification. In 2011, Orhan et al. [6] used the k-means algorithm to cluster from the wavelet coefficients, then classified by a multilayer perceptron neural network (MLPNN). This model yielded maximum accuracy of two-class and three-class classifications that are 100% and 96.67%, respectively. In 2012, Acharya et al. [7] proposed a methodology for the automatic detection of normal, inter-ictal, and ictal from EEG signals. They extracted four entropy features, then fed to a fuzzy classifier. This methodology achieved an accuracy of 98.1%. In 2014, Kaya et al. [8] used the one-dimensional local binary pattern (1-D-LBP) to extract features from raw EEG and, respectively, combined with five different classifiers, such as Bayes Net, SVM, ANN, logistic regression (LR) and functional tree (FT). The best-performing classifier was the Bayes Net classifier, which achieved 99.5% and 95.67% maximum accuracy for two-class and three classifications, respectively. The worst performing classifier was the LR classifier, which gained 96.50% and 66.67% maximum accuracy for two-class and three classifications, respectively. In 2015, Sharma et al. [9]

proposed the features based on the phase space representation for the classification of epileptic seizure and seizure-free EEG signals. They employed the least squares support vector machine as a classifier, which gave 98.67% accuracy. In 2016, Sharmila et al. [10] studied the performance of the 14 different combinations of two-class epilepsy detection. They employed naive Bayes (NB) and k-nearest neighbor (KNN) classifiers for the derived statistical features from DWT, and the NB classifier obtained an accuracy of 100% in the classification of healthy eyes open and epileptic EEG data. In 2017, Zhang et al. [1] employed local mean decomposition (LMD) to decompose raw EEG signals into several product functions (PFs) and then fed the features into five classifiers. The authors reported that the best-performing classifier was the SVM optimized by genetic algorithm (GA-SVM), and the average classification accuracy was equal to or higher than 98.1%. Bhattacharyya et al. [11] computed the Q-based entropy by decomposing the signal with the tunable-Q wavelet transform (TQWT) into the number of sub-bands and estimating K-nearest neighbor entropies (KNNE) from various sub-band cumulatively and used the support vector machine classifier with the wrapper-based feature selection method to be the classifier. This method achieved an accuracy of 100% and 98.6% accuracy of maximum efficiency for two-class and three classifications, respectively. Zahra et al. [12] presented a data-driven approach to classifying five-class EEG classification using the multivariate empirical mode decomposition (MEMD) algorithm. And ANN was employed to be a classifier, which achieved 87.2% accuracy.

These conventional methods for the detection of seizures use hand-engineered techniques to extract features from EEG signals. And, many of these traditional methods show good accuracy for one problem but fail in performing accurately for others [2]. For example, they identify non-seizure and seizure cases (the two-class classification problem) with excellent accuracy but show poor performance for the detection of three-class epilepsy classification. Deep learning is a new research direction of machine learning that automatically learn the inherent laws and features of sample data. As both the available data and computational ability of hardware continue to increase, deep learning has addressed increasingly complex applications with ever-increasing accuracy [13–15]. Automatic detection of epileptic seizures based on deep learning methods received much attention last year. In 2018, Acharya et al. [16] implemented a 13-layer deep convolutional neural network (CNN) algorithm to address the three-class EEG epilepsy classification problem. The proposed CNN architecture includes five convolutional (Conv) layers, five max-pooling layers, and three fully connected (FC) layers. On this three-class detection problem, it achieved an accuracy of 88.67%. Moreover, Ullah et al. [2] proposed an automatic system for epilepsy detection based on an ensemble of pyramidal one-dimensional convolutional neural network models. The core component of the system is a pyramidal one-dimensional convolutional neural network (P-1D-CNN) model, which consists of three main types of layers: Conv, batch normalization (BN), and FC layers. The classification performance of the P-1D-CNN model is not very satisfactory. Hence, the authors introduced the majority-vote (M-V) module in the final stage of the P-1D-CNN model, which significantly improved the performance of the algorithm. In almost all the cases of two-class and three-class concerning epilepsy detection problems, it has given the accuracy of 99.1 ± 0.9%. In 2019, Turk et al. [15] obtained two-dimensional frequency-time scalograms by applying Continuous Wavelet Transform (CWT) to EEG records containing five different classes and used the CNN structure to learn the properties of the scalogram images. On all the two-class, three-class, and five-class classification problems involving seizures, its recognition accuracy is 98.5%–99.5%, 97.0%–99.0%, and 93.6%, respectively. Moreover, Hussein et al. [16] introduced a deep Long Short-Term Memory (LSTM) network to learn the high-level representations of different EEG patterns, using one FC layer to extract the most robust EEG features relevant to epileptic seizures. This model achieved 100% accuracy of the two-class, three-class, and five-class classification problems.

Despite the encouraging seizure detection results gained using the CNN models mentioned above, several improvements can still be achieved. First, some of these CNN models have relatively single model structures. The second issue is the small number of available samples, which is not enough to train a deep neural model. As such, we felt motivated to develop a CNN model for detecting seizures efficiently with raw EEG signals. To address these issues, firstly, we add the BN layer and dropout layer into the convolutional blocks for learning features, which may help in detecting seizures efficiently. Secondly, the segments of raw EEG were divided into many non-overlapping chunks to increase the number of samples for training and test, which may help in using a small amount of available data for more fully training a deep model. Results have shown that the proposed cognitive system is efficient in cognitive computation of EEG signals and also is advantageous in detecting seizures.

2.2 Preliminaries for Cognitive Computation

2.2.1 Description of EEG Dataset

Our seizure recognition experiments are conducted using the widely used and publicly available EEG database produced by Bonn University [17]. This database consists of five diverse subsets (set A-E) denoted as Z, O, N, F, and S. Sets A and B are comprised of surface EEG recordings of healthy volunteers in the wakeful state with eyes open and eyes closed, respectively. On the other hand, Sets C, D, and E are gathered from patients with epilepsy. Thereinto, Sets C and D were recorded during seizure-free intervals. Set C was recorded from the hippocampal formation of the opposite hemisphere of the brain. Set D was recorded from within the epileptogenic zone. Set E only included seizure activities. Each of these sets contains 100 single-channel recordings of EEG signals with a sampling rate of 173.61 Hz and a duration of 23.6 s.

The corresponding time-series is sampled into 4097 data points. Besides, Rochester institute of technology divided every 4097 data points into 23 chunks.

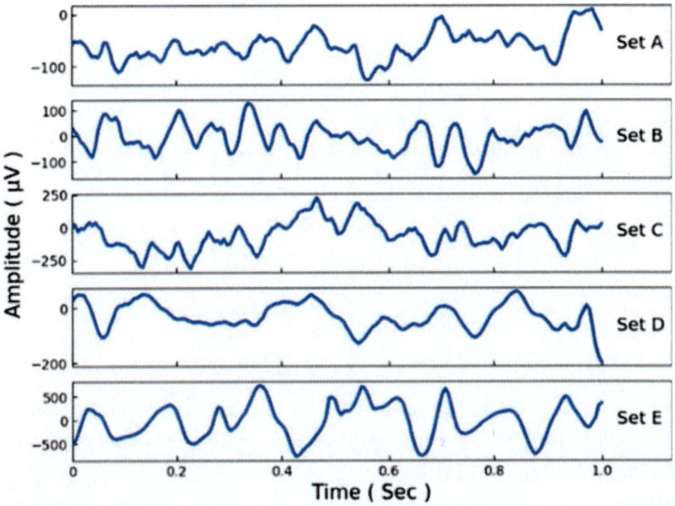

Fig. 2.1 Sample EEG signals in this study

Each chunk contains 178 data points for 1 s (https://archive.ics.uci.edu/ml/datasets/Epileptic+Seizure+Recognition).

To increase the number of samples for training a deep model, the Bonn dataset in this format is adopted, which amount of sample increase 22 times. Therefore, the number of each category has 2300 EEG samples. Sample EEG signals of five EEG classes are shown in Fig. 2.1.

2.2.2 Construction of the Cognitive System

The deep CNN model [18] can automatically learn the features of EEG signals and performs classification in an end-to-end manner. The overall CNN architecture proposed in this paper is shown in Fig. 2.2, which consists of feature extracted stage and classification stage. Firstly, the input one-dimensional raw EEG data is normalized to zero mean and unit variance. The features are extracted through three convolutional blocks, where each block consists of five layers. In detail, the first layer computes multiple convolutions in parallel to generate a set of linear activation responses. The second layer is BN, which is used to solve the internal variable shift. Each linear activation response passes a non-linear activation function in the layer. The activation function used in this work is the Rectified Linear Unit (ReLU) [19]. In the fourth layer, the dropout technology [20] is employed to prevent overfitting. The last layer of the block is the max-pooling layer, which introduces translation invariance. In the structure, the second and third blocks are similar to the first.

At the end of the third convolutional block, the feature maps are flattened into a one-dimensional vector that is connected to the FC layer for integrating features.

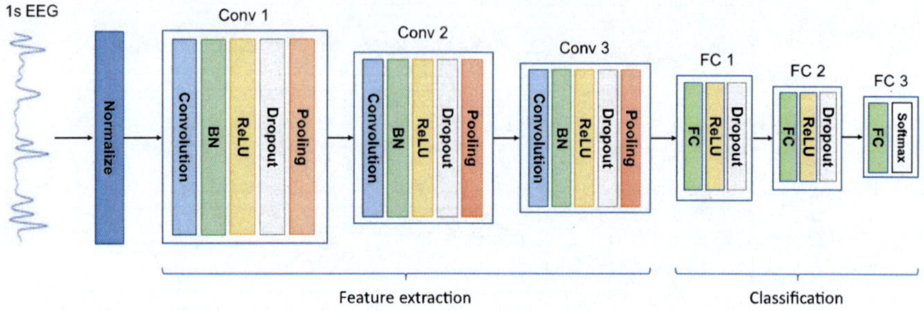

Fig. 2.2 The proposed cognitive system

The first two FC layers employ ReLU as the activation function, followed by a dropout layer. The third FC layer applies softmax as the activation function which will output a vector of probabilities corresponding to each category. To choose better model parameters, we explored eight models with different specifications. Details are described in the experimental results and discussion section. We select model M7. Table 2.1 shows the details of the proposed CNN structure.

2.3 Cognitive Computation Processes

2.3.1 Process of the Convolution Operation

A convolutional neural network (CNN) is a neural network designed to process data with similar network structures. The image can be regarded as a two-dimensional pixel grid. Similarly, time-series data can be considered as a one-dimensional grid formed by regularly sampling on time axis. The convolutional block of conventional CNN includes three layers: convolution, activation function, and pooling. For the one-dimensional EEG data used in this paper, the convolution operation is as follows:

$$s(t) = (x * w)(t) = \sum_{a} x(a)w(t - a) \tag{2.1}$$

Convolution network has the characteristics of sparse interaction. So, it means fewer parameters need to be stored, which not only reduces the storage requirements of the model but also simplifies the calculation. At the same time, the parameters shared by the convolution kernel ensure that we only need to learn parameters that are many orders of magnitude smaller. Convolution is a kind of special linear operation, and activation function brings nonlinear characteristics into the network. The Rectified Linear Unit function (ReLU) is the most commonly used activation function in CNN, which overcomes the vanishing gradient problem, allowing models to learn faster and perform better. Equation (2.2) shows the ReLU function.

Table 2.1 The details of the CNN structure used in this research

Block	Type	Number of neurons (output layer)	Kernel size for each output feature map	Stride
1	Convolution	139 × 20	40	1
	BN	139 × 20	–	–
	ReLU	139 × 20	–	–
	Dropout	139 × 20	–	–
	Max-Pooling	70 × 20	2	2
2	Convolution	51 × 40	20	1
	BN	51 × 40	–	–
	ReLU	51 × 40	–	–
	Dropout	51 × 40	–	–
	Max-Pooling	26 × 40	2	2
3	Convolution	17 × 80	10	1
	BN	17 × 80	–	–
	ReLU	17 × 80	–	–
	Dropout	17 × 80	–	–
	Max-Pooling	9 × 80	2	2
4	FC	64	–	–
	ReLU	64	–	–
	Dropout	64	–	–
5	FC	32	–	–
	ReLU	32	–	–
	Dropout	32	–	–
6	FC	2 or 3 or 5	–	–

$$f(x) = \max\{0, x\} \tag{2.2}$$

The pooling function can reduce the feature maps and reduce the number of parameters used in the network, which replaces the output of the system at a specific position. For example, max-pooling gives the maximum value in several neighborhoods. The pooling can also help to make the representation approximately invariant to small translations of the input.

2.3.2 Layers of Batch Normalization

The CNN proposed in this study adds the BN and dropout layers on the traditional convolutional blocks. When training the deep neural network, the parameters of each layer are closely related to each other. When the parameters of one layer change, it will cause the chain changes of the distribution of other layers, which is called the internal covariate shift. And the internal vary shift makes it difficult for us to choose an appropriate learning rate. To tackle this problem, Ioffe and Szegedy [21] developed BN technology which can almost reparameterize any deep networks, significantly reducing the problem of coordinated updates between multiple layers. The technology takes normalization as part of the model architecture and normalizes each mini-batch.

During training, BN calculates the sample mean and standard deviation for the mini-batch response H in backpropagation by

$$\mu = \frac{1}{m}\sum_{i} H_i \tag{2.3}$$

$$\sigma = \sqrt{\delta + \frac{1}{m}\sum_{i}(H-\mu)_i^2} \tag{2.4}$$

where the delta component δ is kept at a small positive value, and is added only to avoid the gradient becoming undefined where the true standard deviation is zero. And they are used to normalize H by

$$H' = \frac{H-\mu}{\sigma} \tag{2.5}$$

BN is also very useful in accelerating the convergence of the training phase and prevents overfitting. The technology has become a common practice.

Therefore, we employ BN after every convolution layer.

2.3.3 Feature Fusion and Classification

A deep neural network needs to learn a large number of parameters, which is likely to cause over-fitting in the case of a small dataset. To address this issue, the authors [20] developed dropout technology to prevent the co-adaptation of feature detectors. The critical idea of dropout is to randomly drop units with a pre-defined probability (along with their connections) from the neural network during training. It significantly reduces overfitting and gives significant improvements over other regularization methods. In the proposed model, we add the dropout lay after each ReLu activation function.

The output of the last convolutional block represents high-level features in the EEG signals. The fully-connected layer is a usual manner of learning non-linear combinations of these features. All the neurons in the last max-pooling layer are connected with all the neurons of the first FC layer. We used three FC layers. The number of neurons in the final FC layer (FC3) relies on the detection problem, e.g., for the two-class, three-class, and five-class epileptic classification problem, the number of neurons in FC3 is 2, 3 and 5, respectively.

The softmax activation function is a generalization of the binary form of Logistic Regression. It is commonly applied to the last layer of a deep neural network for constituting a categorical distribution over class labels and obtaining the probabilities of each input element belonging to a label. The mathematical expression of the softmax activation function is given in Eq. (2.6), which represent the respective probabilities of the i-th sample belonging to each category under parameter Θ.

$$h_\theta\left(x^{(i)}\right) = \begin{bmatrix} p\left(y^{(i)} = 1 \middle| x^{(i)}; \theta\right) \\ p\left(y^{(i)} = 2 \middle| x^{(i)}; \theta\right) \\ \vdots \\ p\left(y^{(i)} = k \middle| x^{(i)}; \theta\right) \end{bmatrix} = \frac{1}{\sum_{l=1}^{k} e^{\theta_l^T x^{(i)}}} \begin{bmatrix} e^{\theta_1^T x^{(i)}} \\ e^{\theta_2^T x^{(i)}} \\ \vdots \\ e^{\theta_k^T x^{(i)}} \end{bmatrix} \tag{2.6}$$

2.4 Machine Cognition Processes

2.4.1 Training of the Cognitive System

Training the proposed model needs the weight parameters to be learned from the EEG data. For learning these parameters, we employed the conventional back-propagation algorithm with cross-entropy as the loss function. And, we used the stochastic gradient descent method with Adam optimizer that is based on the adaptive estimation of first-order and second-order moments. The hyper-parameters of Adam algorithm are as follow: learning rate (0.0005), beta1 (0.9), beta2 (0.999). The model was implemented in Keras, a powerful deep learning library, which run on top of TensorFlow. The batch size of 100 is chosen in this work, which used for each training update. To compare the performance measure, we trained all the models that present in this work with 300 epochs.

2.4.2 Evaluation of the CNN Model

For evaluation, we adopted well-known performance metrics, such as accuracy (Acc), precision (Pre), sensitivity (Sen), specificity (Spc), F1. Thereinto, accuracy is one of the most commonly used metrics in literature, and it is defined as a ratio between the correctly classified samples to the total number of samples. The definitions of these performance metrics are as follow:

$$Acc = \frac{TP+TN}{TP+TN+FP+FN} \tag{2.7}$$

$$Pre = \frac{TP}{TP + FP} \tag{2.8}$$

$$Sen = \frac{TP}{TP + FN} \tag{2.9}$$

$$Spc = \frac{TN}{FP + TN} \tag{2.10}$$

$$F1 = \frac{2 \times Pre \times Sen}{Pre+Sen} \tag{2.11}$$

where TP (true positive) is the number of abnormal EEG records, which are correctly identified as abnormal; TN (true negative) is the number of normal EEG cases that are correctly predicted as normal; FP (false positive) is the number of normal EEG cases that are predicted as abnormal; and FN (false negative) is the number of abnormal EEG records that are incorrectly classified as normal.

2.4.3 Treatment of the Statistical Uncertainty

To reduce the statistical uncertainty of test error estimation caused by small-scale test datasets, we adopted tenfold cross-validation for evaluation. The 2300 EEG signals of each category are randomly divided into ten non-overlapping fold. During the *i*-th test, the *i*-th fold of the EEG signals are used for testing while the remaining 9 folds are used for training. The metrics reported in the paper are the average performance obtained from ten evaluations.

2.5 Practical Applications of the Cognitive System

2.5.1 Configuration of the Cognitive Models

Datasets are grouped with different combinations for exploring a general classification model, which is classified into two classes (non-seizures and seizures), three categories (normal, inter-ictal and ictal), and five classes (A, B, C, D, and E). To choose better model parameters, we considered eight models with different configurations. We explored models with different parameters, including the size of the receptive field, the number of neurons, and the dropout probability of the FC layer, for comparison.

Taking the five-class classification problem, for example, the experimental results using 10-folds cross-validation are shown in Table 2.2.

Experiments show that within experimental parameters, a larger size of the receptive field and more neurons in the FC layer make the recognition more effective. The dropout probability of 20% in the FC layers is more effective than a rate of 50%.

Therefore, the parameters of the model M7 with the best performance are used for experiments of two-class and three-class classifications with various combinations.

2.5.2 *Models Performance and Evaluation*

A multiple classification problem can be decomposed into multiple binary classification problems. The result of each classification can be listed as a confusion matrix, which reflects the original and predicted label of each category. Table 2.3 shows the confusion matrix and evaluation metrics of classification normal (B) versus preictal (D) versus seizure (E), as well as the overall classification result. All the metrics are over 96.1%, especially the specificity, which is above 98% in each category and the overall classification.

To check the robustness of the proposed model, we tested 20 combinations. The detail of 10-folds cross-validation results is shown in Table 2.4, in which the average accuracy is employed as overall accuracy. The accuracy of the two-class classification varies from 97.6 to 99.5%, which has the best-performing for A versus E and the worst-performing for D versus E. The accuracy of the three-class recognition problem is between 96.7 and 98.1%. Notably, the accuracy is as high as 98.1% for B versus D versus E. The five-class classification problem is more complicated and harder to solve than the two-class and three-class problems but has an advantage in numerous clinical applications, and the proposed model still obtains an overall accuracy of 93.6%. The proposed model is suitable for various classification problems of the Bonn dataset and has a strong generalization ability.

2.5.3 *Comparisons with Other Cognitive Systems*

Numerous approaches have been presented in the literature to automated detect seizure using the Bonn EEG database. Table 2.5 shows the results of the comparison of the recognition rate of this work with them on various classification problems. The binary classification problem is the problem of identifying non-epileptic seizures and seizures. Classification of healthy volunteers and seizures is A versus E, B versus E, and A + B versus E. Due to the significant differences in this classification, the classification results of the various methods that appear in Table 2.5 are generally outstanding, all above 99%. The classification accuracy of inter-ictal and ictal (C vs. E, D vs. E, and C + D vs. E) is slightly lower than the first binary classification. In particular, both D and E are from the epileptogenic zone, therefore it is difficult to

Table 2.2 The configurations of 8 models using 10-folds cross-validation for the A versus B versus C versus D versus E case

Layers	Parameter	M1	M2	M3	M4	M5	M6	M7	M8
conv1	Number of kernels	20	20	20	20	20	20	20	20
	Size of receptive field	5	5	5	5	40	40	40	40
	Dropout rate	0.2	0.2	0.2	0.2	0.2	0.2	0.2	0.2
conv2	Number of kernels	40	40	40	40	40	40	40	40
	Size of receptive field	3	3	3	3	20	20	20	20
	Dropout rate	0.2	0.2	0.2	0.2	0.2	0.2	0.2	0.2
conv3	Number of kernels	80	80	80	80	80	80	80	80
	Size of receptive field	3	3	3	3	10	10	10	10
	Dropout rate	0.2	0.2	0.2	0.2	0.2	0.2	0.2	0.2
Fc1	Number of neurons	32	32	64	64	32	32	64	64
	Dropout rate	0.2	0.5	0.2	0.5	0.2	0.5	0.2	0.5
Fc2	Number of neurons	16	16	32	32	16	16	32	32
	Dropout rate	0.2	0.5	0.2	0.5	0.2	0.5	0.2	0.5
A versus B versus C versus D versus E	*Acc*	90.47	88.61	91.89	91.20	93.37	91.62	**93.55**	92.92
	Sen	75.98	70.86	79.66	77.79	83.33	78.85	83.73	82.30
	Spe	94.00	92.72	94.92	94.45	95.83	94.71	95.93	95.57

Table 2.3 The confusion matrix for the three-class problem (B vs. D. vs E.) across 10-folds

		Predicted			Accuracy	Sensitivity	Specificity	Precision	F1
		Normal	Preictal	Seizure					
Original	Normal	2263	36	1	98.32	98.39	98.28	96.63	97.50
	Preictal	49	2220	31	97.54	96.52	98.04	96.10	96.31
	Seizure	30	54	2216	98.32	96.35	99.30	98.58	97.45
Overall	–	–	–	–	98.06	97.09	98.54	97.10	97.09

Table 2.4 The accuracies (%) of 10-folds cross validation using model M7

	K1	K2	K3	K4	K5	K6	K7	K8	K9	K10	Mean
A versus E	100	99.57	99.57	99.35	99.35	99.57	99.13	99.57	99.35	99.78	99.5
B versus E	99.78	99.13	99.57	98.91	99.13	99.35	98.70	98.70	98.70	99.13	99.1
C versus E	99.35	98.04	98.04	96.96	98.26	97.39	97.39	97.83	98.48	98.48	98.0
D versus E	97.61	98.04	98.26	98.04	97.17	98.04	96.52	97.17	96.52	98.91	97.6
AB versus E	99.57	99.13	99.57	99.57	99.57	99.13	99.57	99.13	99.57	98.99	99.4
AC versus E	99.28	98.70	99.13	98.84	99.13	98.70	98.99	99.57	99.42	98.55	99.0
AD versus E	98.12	97.83	98.41	98.70	98.41	98.41	98.55	98.55	99.13	98.84	98.5
BC versus E	98.70	98.41	97.68	98.55	98.55	98.99	98.84	99.28	99.57	98.26	98.7
BD versus E	97.39	97.10	97.54	98.84	98.26	97.54	98.41	97.97	97.83	97.39	97.8
CD versus E	97.68	97.54	98.41	97.83	98.41	97.25	98.84	98.41	97.97	97.97	98.0
ABC versus E	99.24	98.26	99.24	98.91	98.80	99.02	98.91	99.24	98.91	98.37	98.9
ABD versus E	98.80	98.37	98.80	98.26	98.80	99.35	98.48	97.93	98.15	98.26	98.5
BCD versus E	98.26	97.61	98.59	98.26	98.59	99.24	98.04	98.70	97.93	98.37	98.4
ABCD versus E	98.96	99.22	98.70	98.52	98.35	99.22	98.78	98.61	99.13	98.09	98.8
A versus C versus E	96.04	97.05	97.00	97.39	94.98	97.58	97.00	96.09	96.81	97.39	96.7
A versus D versus E	97.63	97.10	97.54	95.94	97.00	96.67	97.39	97.87	96.81	96.43	97.0
B versus C versus E	97.63	97.97	98.12	97.68	98.36	97.20	97.87	99.03	97.68	97.58	97.9
B versus D versus E	98.35	98.30	98.07	97.49	98.26	97.97	97.20	98.45	98.45	98.06	98.1
AB versus CD versus E	96.70	97.10	97.74	96.43	96.72	97.97	94.96	97.91	96.96	97.25	97.0
A versus B versus C versus D versus E	92.99	94.37	94.00	93.41	93.36	92.73	93.74	93.25	93.74	93.91	93.6

Table 2.5 The comparison between proposed method and other methods using the same dataset

Data sets	Methodology	Study	Acc (%)	Our Acc (%)
A versus E	TFA + ANN	Tzallas et al. 2009	100	99.5
	DWT + Kmeans + MLPNN	Orhan et al. 2011	100	
	1-D-LBP + FT/BN	Kaya et al. 2014	99.5	
	DWT + NB/KNN	Sharmila et al. 2016	100	
	TQWT + KNNE + SVM	Bhattacharyya et al. 2017	100	
	LMD + GA-SVM	Zhang et al. 2017	100	
	CNN + M-V	Ullah et al. 2018	100	
	CWT + CNN	Turk et al. 2019	99.5	
B versus E	DWT + NB/KNN	Sharmila et al. 2016	99.3	99.1
	TQWT + KNNE + SVM	Bhattacharyya et al. 2017	100	
	CNN + M-V	Ullah et al. 2018	99.6	
	CWT + CNN	Turk et al. 2019	99.5	
C versus E	DWT + NB/KNN	Sharmila et al. 2016	99.6	98.0
	TQWT + KNNE + SVM	Bhattacharyya et al. 2017	99.5	
	CNN + M-V	Ullah et al. 2018	99.1	
	CWT + CNN	Turk et al. 2019	98.5	
D versus E	1-D-LBP + FT/BN	Kaya et al. 2014	95.5	97.6
	DWT + NB/KNN	Sharmila et al. 2016	95.6	
	TQWT + KNNE + SVM	Bhattacharyya et al. 2017	98.0	
	LMD + GA-SVM	Zhang et al. 2017	98.1	
	CNN + M-V	Ullah et al. 2018	99.4	
	CWT + CNN	Turk et al. 2019	98.5	
AB versus E	DWT + NB/KNN	Sharmila et al. 2016	99.2	99.4
	CNN + M-V	Ullah et al. 2018	99.7	
CD versus E	1-D-LBP + FT/BN	Kaya et al. 2014	97.0	98.0
	DWT + NB/KNN	Sharmila et al. 2016	98.8	
	CNN + M-V	Ullah et al. 2018	99.7	
ABCD versus E	DWT + Kmeans + MLPNN	Orhan et al. 2011	98.8	98.8
	DWT + NB/KNN	Sharmila et al. 2016	97.1	
	TQWT + KNNE + SVM	Bhattacharyya et al. 2017	99.0	
	LMD + GA-SVM	Zhang et al. 2017	98.9	

(continued)

Table 2.5 (continued)

Data sets	Methodology	Study	Acc (%)	Our Acc (%)
	CNN + M-V	Ullah et al. 2018	99.7	
B versus D versus E	CNN	Acharya et al. 2018	88.7	98.1
	CWT + CNN	Turk et al. 2019	98.0	
AB versus CD versus E	DWT + Kmeans + MLPNN	Orhan et al. 2011	95.6	97.0
	TQWT + KNNE + SVM	Bhattacharyya et al. 2017	98.6	
	LMD + GA-SVM	Zhang et al. 2017	98.4	
	CNN + M-V	Ullah et al. 2018	99.1	
A versus B versus C versus D versus E	TFA + ANN	Tzallas et al. 2009	89.0	93.6
	MEMD + ANN	Zahra et al. 2017	87.2	
	CWT + CNN	Turk et al. 2019	93.6	

distinguish. In the conventional methods of Table 2.5, Zhang et al. [1] obtained the best-performing, which achieved 98.1% accuracy. In CNN-based technology, Ullah [2] employed CNN and the majority-vote module to classify and gain 99.4% accuracy. Turk et al. [15] used CWT and CNN to recognize and achieved 98.5% accuracy. The proposed model of this work just employed CNN and obtain 97.6% accuracy.

The three-class classification problem further subdivides the EEG records to distinguish normal, inter-ictal and ictal EEG. We compared two types of three-class problem (B vs. D vs. E and A + B vs. C + D vs. E). The proposed model also achieved good performance. Especially in the case of B versus D versus E, its performance reaches the best accuracy of 98.1%, which is obviously better than another model [14] based on CNN only. The five-class classification problem is more complicated and harder to classify than the two-class and three-class problems. It addresses the differentiation between EEG epochs belonging to the same class (e.g., sets A and B, which are both normal; sets C and D, which are both inter-ictal). Therefore, in the literature, the relatively little approach was proposed to address these three types of problems at the same time. The proposed CNN model achieved an accuracy of 93.6%, as good as Turk et al. [15], and better than the conventional methods.

The experiment still needs to be implemented in reducing the learning rate, increasing the number of epochs, which will undoubtedly increase the accuracy of epilepsy recognition, but at the same time will cost more time for training. For a limited number of training samples, we can also try to enhance the dataset, which may be useful for the generalization ability of the model. For example, we can divide the 23.6 s EEG data into many overlapping chunks to further increase the number of samples.

The current level of machine intelligence can be summarized as cognitive computation [22, 23]. This chapter was largely re-organized from [24]. A novel model for robust detection seizures has been proposed, which deals with two-class, three-class, and five-class classification problems. The proposed approach has been developed based on the one-dimensional convolutional neural network model, which takes the raw EEG signal as input. To improve the learning ability of the model, the BN and dropout layers have been introduced to the convolution block. This explains how to construct a machine learning system. To address the issue of the small datasets, the EEG has been divided into many non-overlapping chunks for training and test. The experimental result shows that the proposed model performs well on various EEG classification problems.

References

1. T. Zhang, and W. Chen, "LMD Based Features for the Automatic Seizure Detection of EEG Signals Using SVM, " IEEE Transactions on Neural Systems and Rehabilitation Engineering, vol. 25, no. 8, pp. 1100-1108, 2017
2. I. Ullah, M. Hussain, E.U.H. Qazi et al., "An automated system for epilepsy detection using EEG brain signals based on deep learning approach," Expert Systems with Applications, vol.107, pp. 61-71, 2018
3. H. Ocak, " Automatic detection of epileptic seizures in EEG using discrete wavelet transform and approximate entropy," Expert Systems with Applications, vol. 36, no. 2, pp. 2027-2036, 2009
4. A. T. Tzallas, M. G. Tsipouras and D. I. Fotiadis, "Epileptic Seizure Detection in EEGs Using Time–Frequency Analysis, " IEEE Transactions on Information Technology in Biomedicine, vol. 13, no. 5, pp. 703-710, 2009
5. A. Subasi, M.I. Gursoy, "EEG signal classification using PCA, ICA, LDA and support vector machines," Expert Systems with Applications, vol. 37, no. 12, pp. 8659-8666, 2010
6. U. Orhan, M. Hekim, M. Ozer, "EEG signals classification using the K-means clustering and a multilayer perceptron neural network model," Expert Systems with Applications, vol. 38, no. 10, pp. 13475-13481, 2011
7. U.R. Acharya, F. Molinari and S.V. Sree et al., "Automated diagnosis of epileptic EEG using entropies," Biomedical Signal Processing and Control, vol. 7, no. 4, pp. 401-408, 2012
8. Y. Kaya, M. Uyar and R. Tekin et al., "1D-local binary pattern based feature extraction for classification of epileptic EEG signals," Applied Mathematics and Computation, vol. 243, pp. 209-219, 2014
9. R. Sharma, R.B. Pachori, "Classification of epileptic seizures in EEG signals based on phase space representation of intrinsic mode functions," Expert Systems with Applications, vol. 42, no. 3, pp. 1106-1117, 2015
10. A. Sharmila, and P. Geethanjali, "DWT Based Detection of Epileptic Seizure From EEG Signals Using Naive Bayes and k-NN Classifiers", IEEE Access, vol. 4, pp.7716-7727, 2016
11. A. Bhattacharyya, R. Pachori, A. Upadhyay et al., "Tunable-Q wavelet transform based multi-scale entropy measure for automated classification of epileptic EEG signals." applied sciences, vol.7, no.4, 385, 2017
12. A. Zahra, N. Kanwal, N. u. Rehman et al., "Seizure detection from EEG signals using Multivariate Empirical Mode Decomposition," Computers in Biology and Medicine. vol. 88, pp. 132–141, 2017
13. Y. LeCun, Y. Bengio, and G. Hinton, "Deep learning," NATURE, vol. 521, pp. 436–444, 2015
14. C. Huang, Y. Lan, G. Xu et al. "A Deep Segmentation Network of Multi-scale Feature Fusion based on Attention Mechanism for IVOCT Lumen Contour". IEEE/ACM Transactions on Computational Biology and Bioinformatics, 2020
15. M. Li, S. Dong, Z. Gao et al., "Unified model for interpreting multi-view echocardiographic sequences without temporal information". Applied Soft Computing, vol. 88, article 106049, 2020

16. U.R. Acharya, S.L. Oh and Y. Hagiwara et al., "Deep convolutional neural network for the automated detection and diagnosis of seizure using EEG signals," Computers in Biology and Medicine, vol. 100, pp. 270-278, 2018
17. O. Turk, M. S. Ozerdem. "Epilepsy Detection by Using Scalogram Based Convolutional Neural Network from EEG Signals," brain sciences. vol. 9, no.5, article no.115, 2019
18. R. Hussein, H. Palangi and R.K. Ward et al., "Optimized deep neural network architecture for robust detection of epileptic seizures using EEG signals," Clinical Neurophysiology, vol.130, no.1, pp. 25-37, 2019
19. R.G. Andrzejak, K. Lehnertz and F. Mormann et al., "Indications of nonlinear deterministic and finite-dimensional structures in time series of brain electrical activity: Dependence on recording region and brain state," Physical Review E, vol. 64, no. 6, Article ID 061907, 2001
20. Y. LeCun, "Generalization and Network Design Strategies," in Connectionism in perspective, pp. 143–155, Elsevier Zurich, Switzerland, 1989
21. V. Nair and G. E. Hinton, "Rectified linear units improve restricted Boltzmann machines," In Proceedings of the International Conference on Machine Learning (ICML), pp. 807–814, Haifa, Israel, June 2010
22. N. Srivastava, G. Hinton, A. Krizhevsky et al., "Dropout: A simple way to prevent neural networks from overfitting," Journal of Machine Learning Research, vol. 15, no. 1, pp. 1929-1958, 2014
23. S. Ioffe, C. Szegedy. "Batch normalization: accelerating deep network training by reducing internal covariate shift," in Proceedings of the 32nd International Conference on International Conference on Machine Learning, vol. 37, pp. 448–456, Lille, France, July 2015
24. W. Zhao, W. B. Zhao, W. F. Wang, et al. "A Novel Deep Neural Network for Robust Detection of Seizures Using EEG Signals," Computational and Mathematical Methods, Article ID 9689821, 2020

Chapter 3
Data Mining in Environments Sensing

Abstract This chapter aims to explain the processes of machine cognition for a better understanding of environmental changes at the current level of machine intelligence and conjecture how evolution of the machine brain would change the future way of knowledge discovery (data mining) in environments sensing. In order to strengthen the continuity of Chap. 2, we present a parallel framework based on MPI for large dataset to extract power spectrum features of EEG signals so as to improve the speed of brain signal processing. At present, the Welch method has been wildly used to estimate the power spectrum. However, the traditional Welch method takes a lot of time especially for large dataset. In view of this, we added the MPI into the traditional Welch method and developed it into a reusable Master-Slave parallel framework. As long as the EEG data of any format is converted into the text file of a specified format, the power spectrum features can be extracted quickly by this parallel framework. In the proposed parallel framework, the EEG signals recorded by a channel are divided into N overlapping data segments. Then the PSD of N segments are computed by some nodes in parallel. The results are collected and summarized by the master node. The final PSD results of each channel are saved in the text file, which can be read and analyzed by the Microsoft Excel. This framework can be implemented not only on the clusters but also on the desktop computer. In the experiment, we deploy this framework on a desktop computer with a 4-core Intel CPU. It took only a few minutes to extract the power spectrum features from 2.85 GB EEG dataset, seven times faster than using the Python. This framework makes it easy for users, who do not have any parallel programming experience in constructing the parallel algorithms to extract EEG power spectrum. At the end of this chapter, reinforcement learning for solving a blind separation problem is carried out to highlight the necessity to change the way of data mining in environments sensing.

3.1 Introduction

EEG is a recorded signal of electrical activity of the brain, which is collected from the scalp through electrodes. The application of EEG has important practical value in medical treatment, military, sports and the intelligence fields, which has been widely

W. Wang et al., *Interdisciplinary Evolution of the Machine Brain*,
Research on Intelligent Manufacturing,
https://doi.org/10.1007/978-981-33-4244-6_3

recognized by all the researchers. So far, scientists from a various disciplines have achieved good results in this field. For instance, scientists from American Wadsworth Centre help the paralyzed people input 36 characters via signals instead of using their own fingers. These signals correspond to specific brain activity. In China, scientists from Tsinghua University have designed an automatic dialling system. The system is connected to the computer for real-time dialling by interpreting the thinking mode of the brain as corresponding numbers [1–6]. The research of pattern recognition of EEG signal includes the following steps: data collection, data storage, data processing, data classification and recognition as shown in Fig. 3.1.

In order to record EEG signals, several electrodes need to be placed on the scalp. Traditional device usually has 20 electrodes. However, recently EEG devices with as many as 256 electrodes have been used, as shown in Fig. 3.2a. The increase in the number of electrodes enables recording of huge data thereby making the data processing stage in Fig. 3.1 more important and complicated. This not only consumes a lot of computer resources but also leads to poor data extraction thereby directly affecting the accuracy of classification. There are many signal processing methods which can be used to extract EEG features with good discrimination. These methods include time domain analysis, frequency domain analysis, and time-frequency analysis. A comparative analysis of different approaches to spectral signal representation was performed in the paper [7]. These approaches include power spectral density (PSD) techniques, atomic decompositions, time-frequency (t-f) energy distributions, continuous and discrete wavelet approaches, from which band power features can be extracted and used in the framework of motor imagery (MI) classification. It is pointed

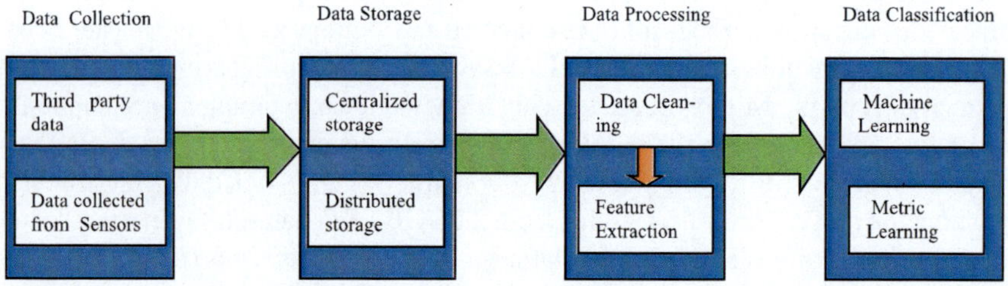

Fig. 3.1 Steps of pattern recognition of EEG signal

Fig. 3.2 **a** 256 electrodes device, **b** wearable EEG device, **c** EEG device for disable

out that among all the feature types of EEG signals, PSD approaches demonstrate to be the most consistent, robust and effective in extracting the distinctive spectral patterns for accurate discrimination between left and right MI induced EEGs.

At present, many methods are used to calculate the PSD of signals. The Welch method is one of the most popular, in which users calculate the PSD of EEG signal in Python or Matlab environment. The function Scipy.signal.welch is used in Python and pwelch is used in Matlab. In case of small amount of data, PSD of EEG signal can be quickly obtained by those two functions. But from Fig. 3.2b, c, we can see that with rapid developments in science and information technology and the up-coming of 5G era, wearable devices will widely be preferred. EEG signals of human activities can be collected by these devices. We can collect current working or learning status of some specific people (such as drivers, students, etc.) via analysis of their EEG signals [8, 9]. If we calculate the PSD in the Matlab or Python environment, it will take a long time which we cannot bear it. Therefore, we focus on how to calculate the PSD of EEG signals by parallel approaches.

Parallel approaches can improve processing speeds but will need technologies that support distributed computations. Nowadays, there are two commonly different frameworks for big data analysis. One is Apache Spark and the other is OpenMP/MPI. The two computational frameworks are compared and analyzed in literature [10]. Apache Spark has the advantage of good data management. OpenMP/MPI is faster than Apache Spark by an order of magnitude. Motivated by this analysis, we present a parallel framework for large dataset to extract power spectrum features of EEG signals, which can be implemented in Linux and MPI environment.

The main contributions of this paper are threefold:

(1) According to the principle of Welch algorithm, we propose a parallel framework of Welch algorithm PFwelch to compute PSD of EEG. The architecture of the PFwelch is based on master-slave mode. The EEG signals recorded by each channel are divided into N overlapping data segments. Then the N segments are computed by master and slave node in parallel. The results are collected and summarized by the master node. The final Welch PSD results of each channel is saved in the text file, which can be read and analyzed by the Microsoft Excel.
(2) A middle file of specified format was used. There are many kinds of EEG datasets in the world, different datasets have different file formats. In order to process any kinds of EEG data by the proposed parallel framework, EEG data need to be converted into middle files. The relationship between the middle file and the parallel framework is depicted in Fig. 3.3.
(3) A comparative experiment was designed. In the experimental stage, we first run the function pwelch in the Matlab environment to extract PSD features from the EEG signal as a baseline for comparison and subsequently run the PFwelch on the Ubuntu platform. The results show that the PFwelch have the same result as function pwelch. After this, we run the Python function Scipy.signal.welch in the same environment with PFwelch. The experimental result shows that the proposed parallel framework is 7 times faster than using Python.

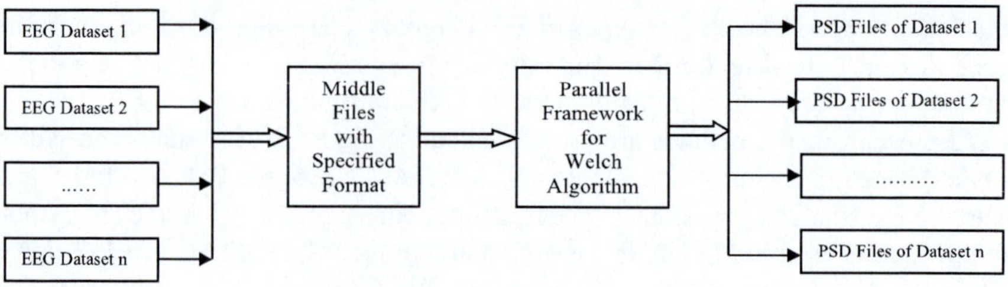

Fig. 3.3 Relationship between the middle file and the parallel framework

This paper is organized as follows. Section 3.2 takes a brief overview of the principle of Welch method. Then presents a serial algorithm of the Welch method and the proposed parallel framework of Welch method. In Sect. 3.3, we present experimental results and analysis. Discussion and conclusions are presented in Sect. 3.4.

3.2 A Parallel Framework for Feature Extraction

3.2.1 Welch's Method

The power spectral density (PSD) exhibits how the power is contained in a signal in the frequency domain. Welch's method and the multi-taper approach have shown the best performance among the PSD estimators [11]. The Welch algorithm [12] is exhibited in Fig. 3.4.

From the Fig. 3.4, we can demonstrate the Welch algorithm in the following mathematical form:

The input signal $x[n]$, $n = 0, 1, \ldots N - 1$, is split into many overlapping segments. In most cases, a overlap of 50% is applied when the input signal is divided into segments. Let the length of each segment be L, the total number of segments be N_s. The formula for the data in the ith segment is as follows:

$$x_i = x[i \times L/2 + n], \text{ where } n = 0, \ldots, L - 1, \; i = 0, 1, 2, \ldots, N - 1 \quad (3.1)$$

The procedure of the segmentation is illustrated in Fig. 3.5.

The relationship of the sampling length N, number of overlapping points N_D, number of segments N_S and segmental length L is

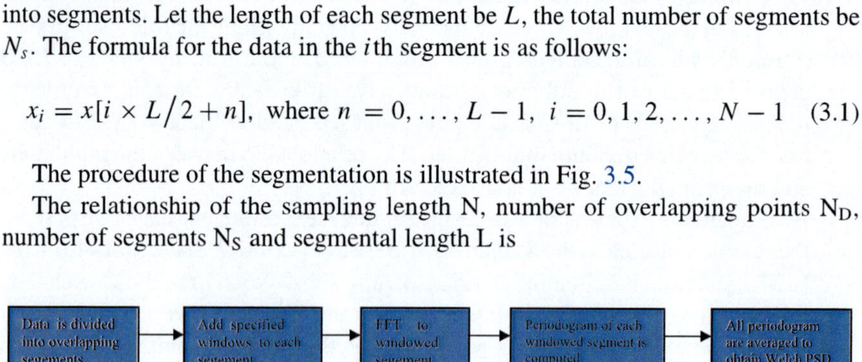

Fig. 3.4 Welch PSD algorithm

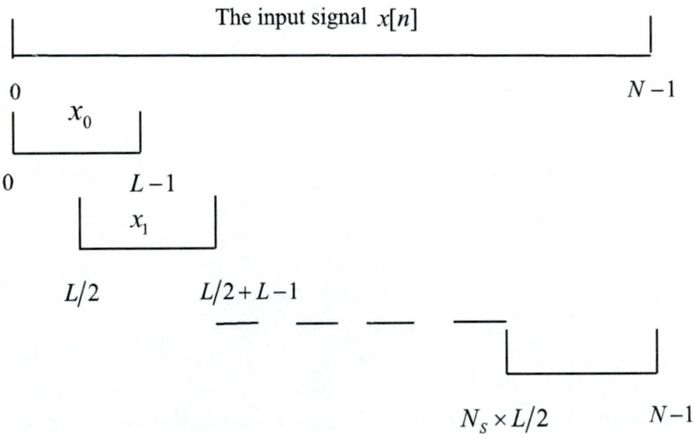

Fig. 3.5 Illustration of signal segmentation

$$N = L + (L - N_D)(N_s - 1) \tag{3.2}$$

A smooth window, $w(n)$ is applied to each segment. Generally, we usually use Hamming window. The formula of Hamming window for each segment is as follows:

$$w(n) = 0.54 - 0.46\cos[2n\pi / L] \tag{3.3}$$

where $n = 0, 1, 2 \ldots, L - 1$, L denotes the length of each segment. Figure 3.6 displays a 256-point Hamming window in the time domains and frequency domain with Matlab.

The purpose of the window function is to prevent the spectral leakage [13]. Figure 3.7a shows the spectrum leakage of the original signal. Figure 3.7b exhibits that the spectrum leakage can effectively be reduced by Hamming window.

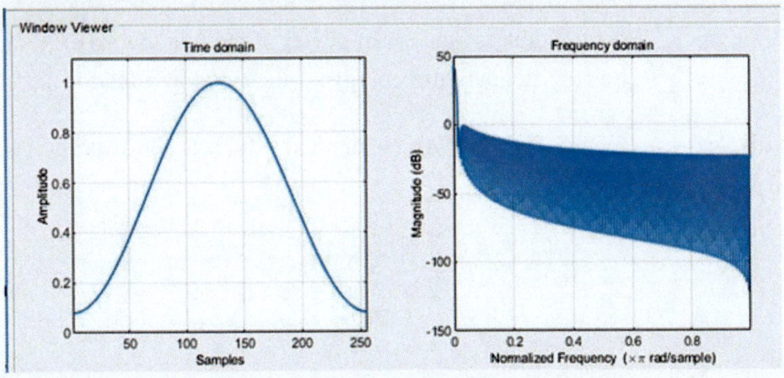

Fig. 3.6 Hamming window in the time and frequency domain with Matlab

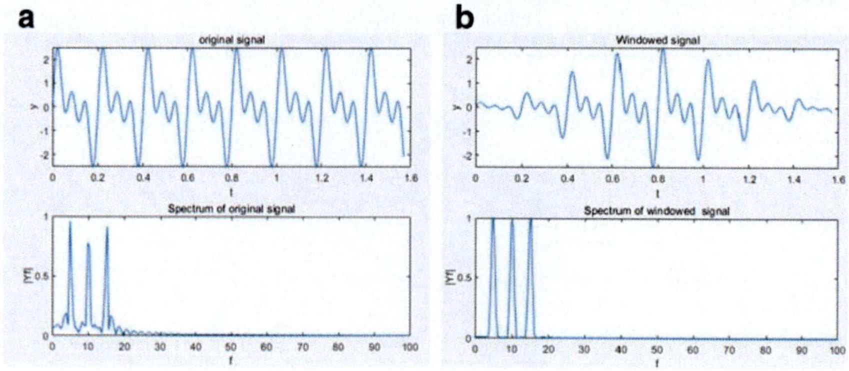

Fig. 3.7 **a** The spectrum leakage of the original signal, **b** windowed signal

From the formula (3.3), we can get the formula (3.4) for ith segment of data after being windowed

$$W_i = x_i(n) \times w(n) \tag{3.4}$$

Fourier transform of each windowed segment is computed. The formula is (3.5):

$$A_i(k) = x_i(n)w(n)e^{-j\frac{2\pi}{N}nk} \tag{3.5}$$

where A_i is the Fourier transform result of the ith windowed segment, $i = 0, 1, \ldots, L-1$.

The periodogram of each windowed segment is computed by using following formula:

$$\phi_i = \frac{1}{LU}|A_i(k)|^2 \tag{3.6}$$

where $U = \frac{1}{L}\sum_{n=0}^{L-1} w^2(n)$ denotes the mean power of the window $w(n)$.

So, $LU = \sum_{n=0}^{L-1} w^2(n)$ denotes the energy of the window function $w(n)$ with length L.

Finally, we can get the PSD by Welch method which is the average of those periodograms, i.e.,

$$S(k) = \frac{1}{L}\sum_{i=0}^{L-1} \phi_l(k) \tag{3.7}$$

3.2.2 *Serial Algorithm of Welch Method*

In order to design a good parallel program, it is necessary to understand the traditional serial algorithm. According to the above description of Welch method, the serial algorithm of the Welch method is written as follows:

Algorithm 1 Serial algorithm of the Welch method

```
Input :
   EEG_file: EEG Signal File
   welch_len: the length you want to welch
   seg_len: length of each segment
   Fs: sampling rate
Output:
   Wel_psd : the Welch estimate of PSD
1. Initialization: over_len= seg_len/2; // Length of overlap between the successive segments
2 Data are read from EEG_file and stored into the arrary welch[n];
3. Calculate how many segments : n_segs
4. for index in range (0, seg_len)
5.     Create Hamming window win[i];
6.     u += win[index]*win[index];// Compute the accumulated value of u;
7.     index++;
8. end for
9. for seg_start in range (1 ,x_len-seg_len+1 )
10      seg_end = seg_start+seg_len -1;
       // seg_start and seg_end is start point and end point of each data segment, respectively.
11      apply Hamming window to the data of welch [n] between [seg_start, seg_end];
12      Do FFT for windowed data. The result is stored in variable fft_x;
13     Calculating the square of the amplitude of the fft_x .The result is stored in variable pgram;
14     Summate power spectra Pxy += pgram;
15     start_seg +=seg_len-overlap;
16.end for
17 Wel_psd =Pxy/( n_fftsxFsxu);
```

3.3 Proposed Parallel Framework of Welch Method

3.3.1 *Program Structure*

According the steps of Serial Algorithms of Welch Method in 2.2, it can be seen that this algorithm can be implemented in parallel with MPI. The structure of parallel algorithm is Master-Slave and is demonstrated as Fig. 3.8.

The structure of PFWelch algorithm in Sect. 3.2 contains following stages: input, split, map, reduce and output stage. One channel data is split into seven segments, which are allocated to four nodes according to certain rules. All nodes are responsible for computing. In the reduce stage, the master process is responsible for receiving the result from slave processes and compute the final PSD. Since the parallel implementation of fast Fourier transform has been maturely developed [14–20], we need not give the relevant details. The parallel algorithm of the Welch method can be described as follows:

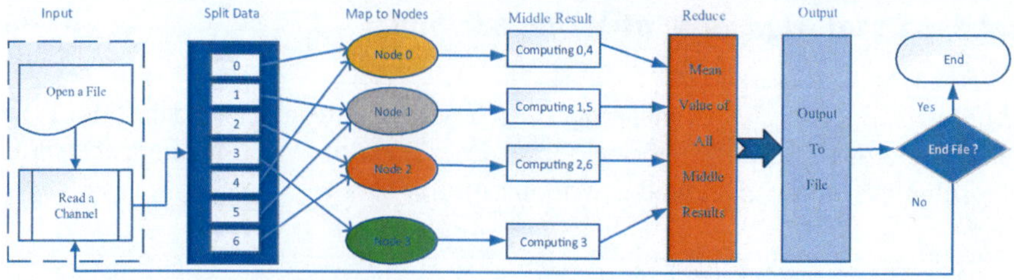

Fig. 3.8 Program structure of PFWelch

Algorithm 2 Parallel algorithm of the Welch method

Input :
  EEG_file: EEG Signal File
  welch_len: the length you want to welch
  seg_len: length of each segment
  Fs: sampling rate
Output:
  Wel_psd : the Welch estimate of PSD
Begin
 1. Initialization: MPI_Init;
   rank ←MPI_Comm_rank;
   size←MPI_Comm_size ;
   overlap_len= seg_len/2;
 2. Read a EEG_file
 3. Data from EEG_file are stored into the array welch[n];
 4 Calculate how many segments : n_segs
 5 for node = 1 to size par-do // Parallel process
 6 apply Hamming window to welch [n]
 // n between [startPos, stopPos]; startPos and stopPos is starting point and end point of certain data segment, respectively.
 7 Do FFT for windowed data. The result is stored in variable fft_x;
 8. Calculating the square of the amplitude of the fft_x .The result is stored in variable pgram;
 9. Summate power spectra Pxy += pgram;
 10. end for
 11. if (rank == 0) //master process
 12. Receive Pxy from each process and compute average of all periodogram.
 14. else //slave processes
 14. Send Pxy to master process
END

3.3.2 *Distribution of Tasks*

The key point of parallel program is how to cooperate with each node [21]. How to distribute tasks evenly to each computing node is one of the major factors affecting the performance of parallel program. From Fig. 3.8, the EEG data of each channel is divided into 7 segments (n_segs = 7). The length of each segment is 64. The allocation scheme is shown in Table 3.1, when the number of nodes is 4 (size = 4).

According to Table 3.1, it is convenient to calculate the signal range that each processor needs to compute. For example, processor 0 needs to handle segments 0

Table 3.1 Task allocation

Label of segment	0	1	2	3	4	5	6
Label of nodes	0	1	2	3	0	1	2

and 4. From formula (3.1), it can be calculated that the signal range is [0, 63] and [128, 191] respectively.

3.4 Experimental Results and Analysis

3.4.1 Description of the Dataset

In the experiment, the data were downloaded from the web site http://kdd.ics.uci.edu/databases/eeg/eeg.html. This dataset arises from a large study to examine EEG correlates of genetic predisposition to alcoholism. It includes measurements from 64 electrodes placed on the scalp sampled at 256 Hz. There are three versions of the EEG data set, small data set, large data set, and full data set. The full data set contains all 120 trials for 122 subjects. The entire set of data is compressed to about 700 MB [22]. When it was uncompressed, its size is approximately 2.8 GB. After all files were converted into middle files, there were in all 11,058 files.

3.4.2 Testing Method and Environment

There are three different performance needs to be tested:

1. The accuracy. The result of MATLAB is compared with that of the parallel framework.
2. The speed. The running time of Python program is compared with that of parallel framework.
3. The speedup. The parallel framework was run with different nodes. The running time was recorded and speedup can be obtained.

The testing environment of hardware is shown in Table 3.2, and the testing environment of software is shown in Table 3.3.

Table 3.2 Testing environment of hardware

Nodes	CPU	Memory	Type
4	Intel(R) Core(TM) i5-4460	8 GB	Desktop computer

Table 3.3 Testing environment of software

Operating System	Parallel environment	Matlab
Ubuntu 16.04	C + Mpich 3.3.1 on Ubuntu	R2018a On Win10

3.4.3 Experimental Results

PSD computation of one EEG channel was conducted in the Matlab by using the Welch function, i.e. pwelch(data,hamming(64),32,64,256). The pwelch parameters mean that sample data is split up into segments, each segment has 64 data points, the overlap of neighboring segments is 32, Hamming window is applied, the sampling rate is 256 Hz. Figure 3.9 shows the result of Welch method in three different window functions in Matlab.

We performed same test in the parallel framework. The PSD results were stored in the text file which can be accessed by Microsoft Excel and is shown in Fig. 3.10.

The leftmost side of Fig. 3.10 is the frequency value, and then the next three columns are the PSD values under three different window functions. Comparison between Figs. 3.9 and 3.10 indicates that the results obtained by both Matlab and PFWelch are consistent, which corroborates the correctness of PFWelch. To test the time performance, we first used different number of nodes in different segments to calculate the PSD of one EEG file with PFWelch (In this experiment, we regard different CPU cores as different nodes). Time consumption (measured in seconds) is recorded in Table 3.4. For comparison, the time cost in Python environment by serial algorithm is also listed in the Table 3.4. Since we haven't Matlab which can be used in Linux, so we haven't carried out such comparative experiment.

Second, we use different number of nodes with 7 segments to calculate the PSD of all EEG files with PFWelch. The time cost is recorded in Table 3.5.

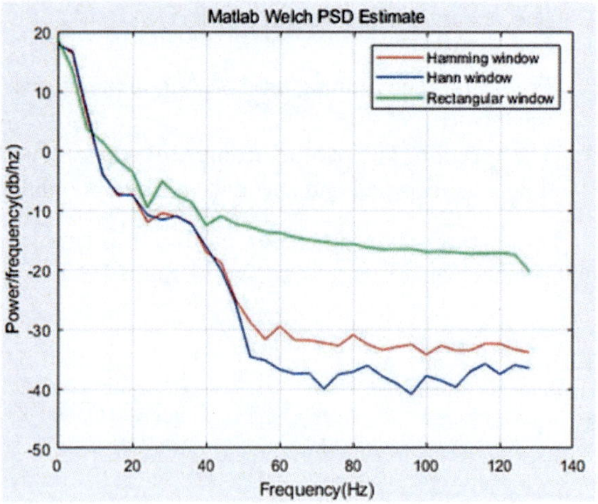

Fig. 3.9 Welch result in matlab

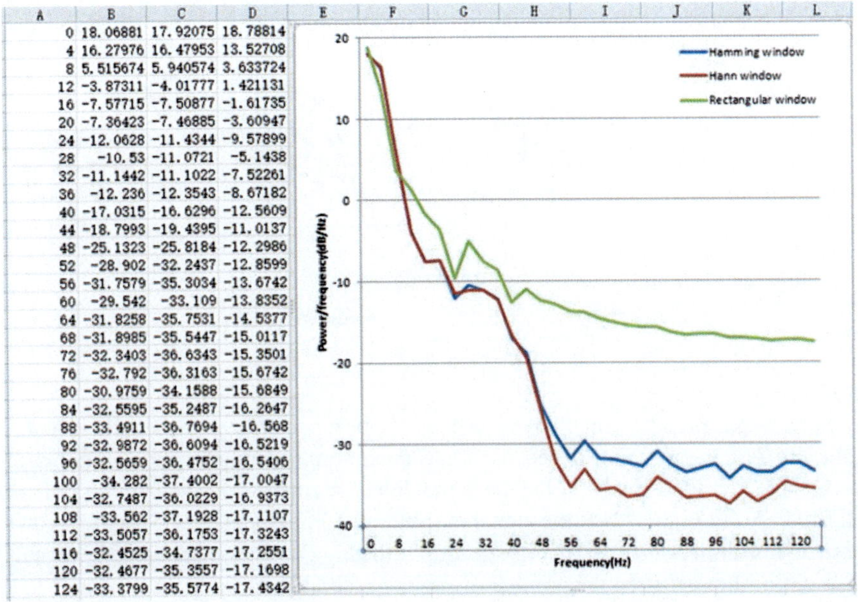

Fig. 3.10 Result of PFWelch

Table 3.4 Time cost of one EEG file

Segments	Nodes				
	1	2	3	4	Python
7	0.0543	0.0413	0.0382	0.0365	0.228
3	0.075	0.06	0.056	0.056	0.235
15	0.064	0.054	0.05	0.047	0.224

Table 3.5 Time cost of all EEG files

Nodes	1	2	3	4	Python
Time cost (s)	478.778	354.008	342.715	337.966	2451

Speedup can evaluate the time performance of PFWelch. The definition of Speedup is: Speedup = Ts/Tp where, Ts indicates the time of serial operations and Tp indicates the time of parallel operations. The Speedups are shown in Fig. 3.11. Figure 3.11 demonstrates that as the number of nodes increase, the Speedup also increases. But in different cases, the increase is not the same. For all EEG files, Fig. 3.11 shows that as the number of nodes increases to a certain value, the speedup slowly increases. The reason is the time it takes includes files open and close. This time cannot be changed when the number of nodes is increased.

Fig. 3.11 Speedup of PFWelch

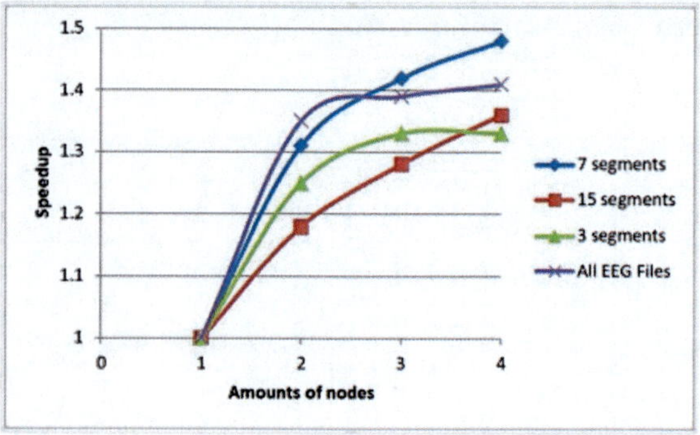

For a single EEG file, the PFWelch shows the best performance when the signal is split into 7 Segments. This is because it has maximum speedup when the number of nodes is 4. Therefore, we divided the signal into 7 segments to calculate the PSD of all EEG files in Fig. 3.11. If the signal is split into 15 Segments, the Speedup grows basically in a linear manner, but its speedup is the lowest. Figure 3.12 clearly reveals the relationship between the number of segments and the speedup.

It can be seen that by increasing the number of nodes, the speedup is improved. But merely increasing the number of nodes cannot improve the speedup. From the Fig. 3.12, we can see that the speedup decreases when the number of segments increases to 15. When the number of segments is 7 and the number of nodes is 4, we get the best speedup performance.

At present, although there are several methods which can extract features from EEG signal, the PSD is still one of the most important methods. But while dealing with a large number of EEG data, it is necessary to improve the calculation speed of PSD. Therefore, the parallel framework proposed in this paper is used to solve the problem of taking long time to extract PSD features from EEG dataset in big data environment. The framework is based on C+ MPI language and is completed by master-slave

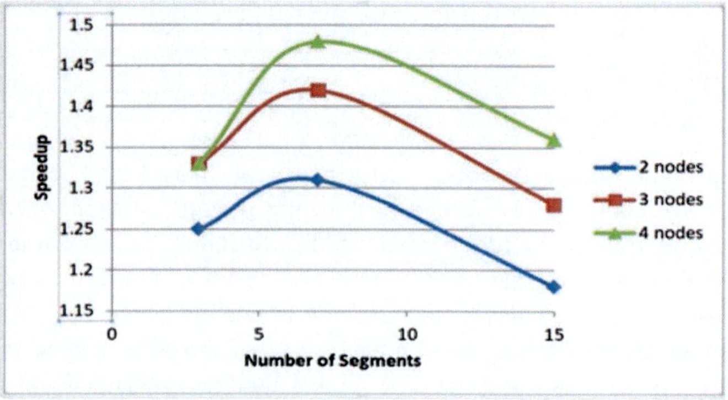

Fig. 3.12 Relationship between the number of segments and speedup

mode. Compared with the traditional serial Welch method, this framework divides the signal into N segments and distributes them evenly to different nodes on which PSD can be calculated in parallel. In the experiment, the values of N are 3, 7 and 15 respectively. Number of nodes is from 1 to 4. For the given data set, we found that although the speedup can be improved by increasing the nodes, but the speedup performance will decrease if there are too many segments. Experiments show that the best performance of the parallel framework can be achieved only when the number of nodes and segments are reasonably selected. The speed of PFWelch is 7 times faster than using Python in the same hardware and operating system platform.

Because of the powerful function of MPI, this framework can be deployed not only on the cluster but also on the desktop computer, which is very convenient for the users. The experimental results also coroborate that the framework is correct, efficient and has a good practical value. It can be applied to extract all kinds of EEG datasets with a little modification. Researchers who are interested in PFWelch can download the source code from https://github.com/abcxq.

3.5 Data Mining in Environments Sensing

Reinforcement learning have been widely recognized as a novel learning system of great significance. A series of critical applications proved this although, until now, its applications in blind separation problem have not been addressed. This study aims to make a first attempt. A theoretical framework for better understanding the inorganic components of soil respiration in arid land is built from a model developed from reinforcement learning. Based on theoretical analyses of rewards evolution in reinforcement learning, regional retrievals from reinforcement learning and unresolved issues beyond reinforcement learning, a "blind C signal separation" (BCS) problem is also presented as a further interpretation of the Rio measurement problem.

3.5.1 Description of the Considered Problem

Problem of partitioning soil respiration (SR) was raised in a long-sought answer to the mysterious carbon (C) missing sink problem, which remains unresolved after many investigations on forest C sink, soil C storage and even, the river and groundwater C transport [23–31]. There still remain considerable uncertainties as to an explicit location of the long-sought "missing sink" and the processes potentially contributed to it, partly because that IoT sensing of atmospheric CO_2 flux is still mainly based on micrometeorological measurements of net ecosystem exchange of CO_2 (NEE) within a FLUXNET community, where NEE was interpreted as the sum of photosynthetic uptake and ecosystem CO_2 fluxs [32]. Recent publications highlighted necessity to take into account abiotic components and introduce a new term *net ecosystem carbon balance* (NECB) [30, 33–43]. A series of "anomalous" CO_2 fluxes over carbonate ecosystems further revealed necessity to partition SR as a sum of soil inorganic respiration (Rio) and soil organic respiration (Ro) [44–47].

In consideration of its possibly, huge effects, it is imperative to determine the representativeness of Rio and presents some further evidence for its contribution to NECB [48–52]. Moreover, its controls are still far from certain and some recent publications suggested that predominant processes involved in Rio might be largely driven by climate change [52–57]. The optimal solution is to develop an analyzer for direct sensing of Rio, which demands a better understanding of spatio-temporal variability in Rio and other components of SR [52, 57]. The original data sources of Rio are collected from a rather laborious and expensive method-soil sterilization, which is almost infeasible on large scales [48, 52]. Time lag between Rio and Rio was detected by the explicit partitioning and reconciling of SR and a semi-experimental method was recently proposed for a direct computation of Rio from SR [57, 58]. These previous studies presented an approximated approach to the overall contributions of Rio in SR and under these background, a comprehensive characterization of the ecological periods of Rio and the spatio-temporal variability of Rio and other SR components can be further established.

This section aims to develop a further partition of Rio and Ro by constructing a model through reinforcement learning of the controls of subterranean CO_2 concentration. Based on theoretical analyses of rewards evolution in reinforcement learning, regional retrievals from reinforcement learning and unresolved issues beyond reinforcement learning, a "blind C signal separation" (BCS) problem is also presented as a further interpretation of the Rio measurement problem.

3.5.2 Problem Formulation and Learning Processes

Based on the previous studies, we argue that there is a hidden loop in subterranean cycles driving CO_2 sequestration from the atmosphere to the soil-groundwater system, especially in the arid regions with saline and sodic soils. Such CO_2 sequestration can be partly interpreted as a rapid carbon mineralization for permanent disposal of anthropogenic carbon dioxide emissions [43].

Consequently, the hypothesized hidden loop is also making a potential contribution to urban CO_2 mitigation within these regions. For the convenience to understand the learning process, we formulate CO_2 sequestration process in the soil-groundwater system as five stages:

(1) CO_2 in the first soil layer reacts with dew and becomes more dissolvable inorganic carbon;
(2) when raining, snow melting or irrigating, these dissolvable inorganic carbon will be partly dissolved and goes into the second layer of the soil-groundwater system;
(3) in more deep layer, i.e., the third layer, since the soils are wet and alkaline, a part of soil organic carbon is converting to dissolvable inorganic carbon;

(4) in the very deep layer, i.e., the fourth layer, most of the dissolvable inorganic carbon are dissolved since it is near the alkaline groundwater;
(5) in the groundwater layer, all the dissolved inorganic carbon migrate into the groundwater.

It is worthy to point out that the hypothesized five stages of CO_2 sequestration in the soil-groundwater system might be simplified if the soil texture is simple. On the contrary, complicated soil texture will also complicate the processes of CO_2 sequestration in the system.

Formation of the blind separation problem is based on a schematic diagram of these five stages, along with how CO_2 sequestration contribute to CO_2 mitigation, is illustrated in Fig. 3.13.

The blind separation problem is how to separate components in SR associated with the hidden loop (termed as soil inorganic respiration R_{io}) and learn their overall significance. Reinforcement learning will be utilized to validate the hypothesis and for a first attempt to solve the blind separation problem, we try to confirm hypothesized controls of different sub-processes in different layers. Generally, CO_2 concentration in the atmosphere is less than that in the soil. But the overall CO_2 emission from saline and sodic soils over arid regions is slight [44], while CO_2 concentration in the atmosphere are significantly increasing due to global warming, industrial CO_2 emission, land-use change in urbanization, plants and animals emission and soil organic respiration. Results in two sub-figures of Fig. 3.14 indicate variations of soil water content in a typical arid region with alkaline soils. When the soil depth increases from 0 cm to 300 cm, soil water content significantly increases (from 0

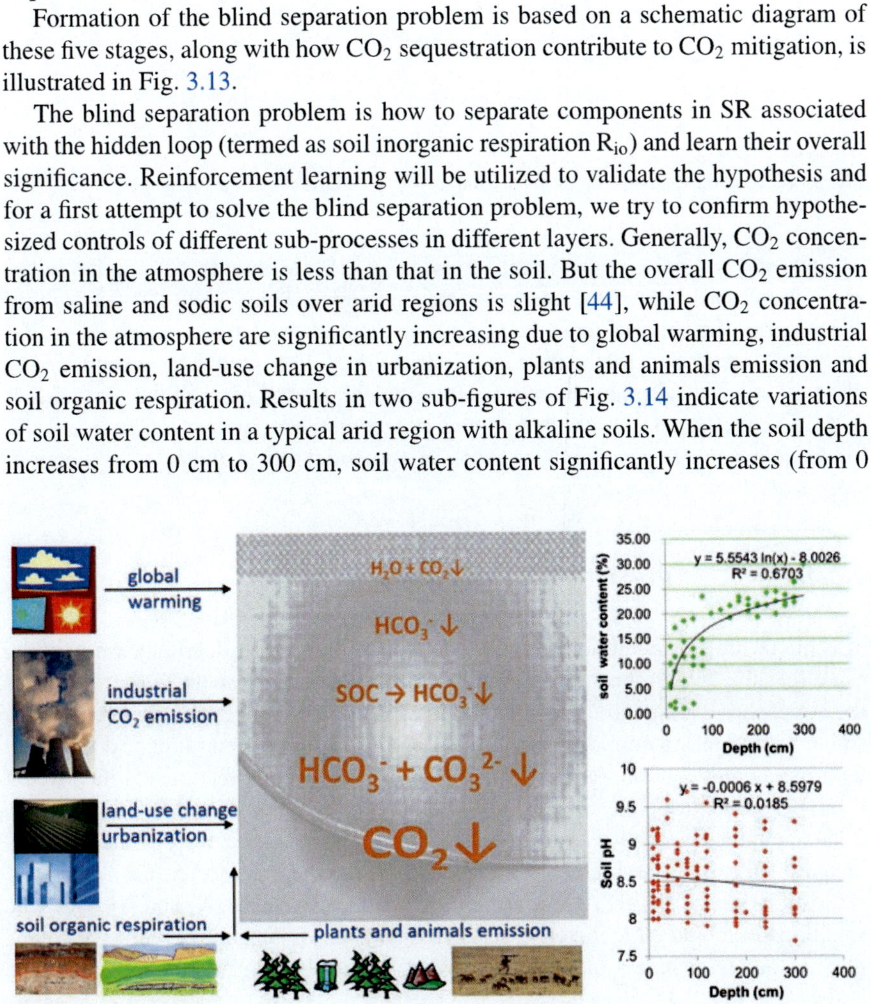

Fig. 3.13 Formation of the blind separation problem, taking into account the role of the hidden loop in urban CO_2 mitigation and decomposition of soil organic carbon (SOC), where raw pH data were collected from [44]. Soils in depth 1–3 m are so dry that there is no a very evident change in pH within the depth range

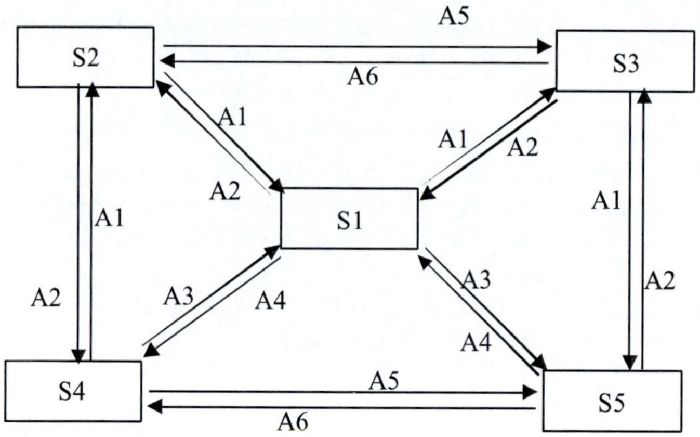

Fig. 3.14 Relationship between the status and actions in reinforcement learning

to 30%) and soil pH is stable in the interval [7.5, 10]. Moreover, CO_2 stored in the soil will be partly dissolved and goes into deep layer when raining, snow melting or irrigating. Because of these facts, CO_2 concentration in the atmosphere can be temporally larger than that in the soil. Ability of the soil-groundwater system in CO_2 dissolution and absorption are getting stronger with the increases of the depth. Most of these sequestrated CO_2 can be finally dissolved in the groundwater. Therefore, the hidden loop in the subterranean cycles plays a potential role in urban CO_2 mitigation.

3.5.3 Process of Reinforcement Learning

To smooth rewards evolution in the process of reinforcement learning, we hypothesize that the subterranean CO_2 concentration is increasing with the increases of the atmospheric temperature and pressure, wind speed, rainfall, CO_2 concentration, soil temperature and humidity and the subterranean CO_2 concentration is decreasing with the increases of atmospheric humidity, rainfall, wind direction, salinity, water level, pH value. Principles in our reinforcement learning are shown in Fig. 3.14.

S_1–S_5 indicate the five states, where S_1 is the starting state, S_3 is the medium increase state, S_5 is the high increase state, S_2 is the medium decrease state, and S_4 is the high decrease state. A_1–A_6 are the six actions, where A_1, A_3 and A_5 represent small, medium and large increase, respectively, and A_2, A_4 and A_6 represent small, medium and large reduction, respectively. Process model for reinforcement learning is shown in Fig. 3.15.

The increase of Q value is calculated from the formula

$$delta = \mathrm{R} + \gamma * \max\{\ \mathrm{Q}(S', \mathrm{A}')\} - \mathrm{Q}(S, A)$$

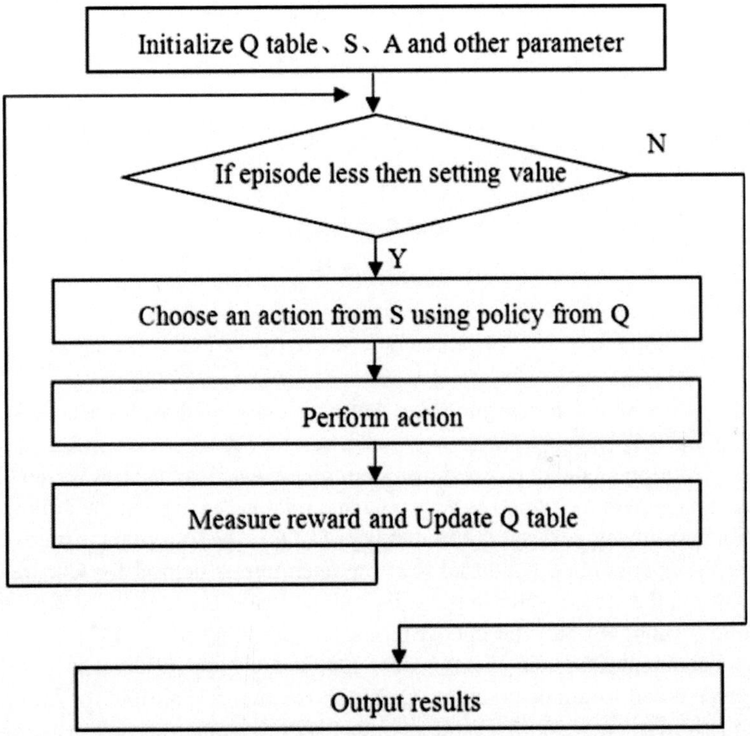

Fig. 3.15 Process model for reinforcement learning employed in the present study

The major principle for updating Q values is

$$Q(S, A) \leftarrow Q(S, A) + \alpha * delta$$

The convergence principle is regular:

(1) the learning process will stop when two Q values are equal or of little difference;
(2) the learning process will stop when attaining the maximal episode.

3.5.4 Results Discussion and Uncertainty Analyses

Based on our hypothesis, the state transition matrix is constructed as

$$\begin{aligned} ST = [&1, 2, 4, 1, 3, 1;\\ &1, 3, 5, 5, 1, 2;\\ &1, 4, 5, 4, 1, 3;\\ &1, 3, 5, 1, 5, 2;\end{aligned}$$

$$1, 4, 2, 3, 1, 5;$$
$$1, 4, 5, 3, 2, 1].$$

The initial rewards associated with the state transition matrix are given as

$$\begin{aligned} R = [&0, -1, 1, -2, 2; \\ &1, 0, 3, -1, 0; \\ &-1, -3, 0, 0, 2; \\ &2, 1, 0, 0, 3; \\ &-2, 0, -1, -3, 0]. \end{aligned}$$

Rewards evolution in reinforcement learning associated with each episode is shown in Fig. 3.16.

The separation model developed from reinforcement learning is performed in the Central Asia, where the winter is very cold and relatively long and the atmospheric conditions are advantageous for the solubility of CO_2 in the soil-groundwater system. Even in the hot summer, nocturnal lower temperature is helpful for CO_2 reacting with dew and the stored water in the soils and motivates the formation of dissolved inorganic carbon. Reanalyzed data from an incorporating model [42] will further expand insights into the role of this hidden loop in CO_2 mitigation, where soil CO_2 flux is reconciled a sum of organic CO_2 efflux (mainly responsible for the surface CO_2 emission) and inorganic CO_2 influx (induced by the subterranean sink). Regional retrievals of soil inorganic respiration in the Central Asia with saline and sodic soils are presented in Fig. 3.17 to further show the significance of the hypothesized hidden loop in CO_2 mitigation.

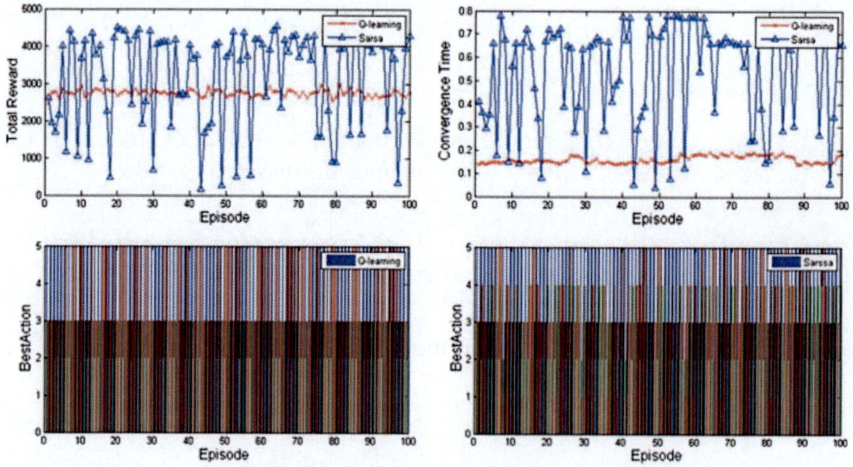

Fig. 3.16 Rewards evolution in reinforcement learning associated with 100 episodes

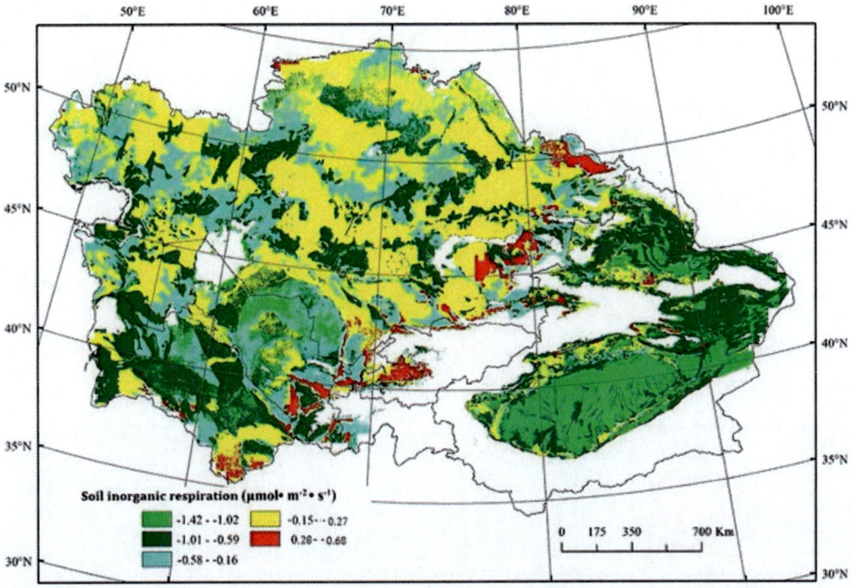

Fig. 3.17 Regional retrievals of soil inorganic respiration from reinforcement learning in the Central Asia

This show that soil in arid regions can temporally acts as a CO_2 sink [39, 41]. Neglecting the anonymous CO_2 sequestration will obscure a significant fraction that contributes to the modern carbon cycle and induce erroneous or misleading conclusions in predicting the future feedbacks in the coupled carbon-climate system [38–41]. Further experiments are worthy to be conducted to further explain the five stages of CO_2 sequestration in the soil-groundwater system.

In previous publications, it was demonstrated that the variations of Rio originate from the physical forcing of abiotic factors such as soils salinity (EC), alkalinity (pH), temperature (TS), and water content (WCS) and their linear relationships with its daily mean intensity appear to be valid within a seasonal cycle as a whole, and the hourly-scale variations of Rio revealed a replication (or a period character), which must be explicitly taken into account in simulations to cover diurnal cycles [48, 52, 57]. Analyses in these studies are based on the database acquired from Fukang Station of Desert Ecology, which is located at the southern periphery of the Gubantonggut Desert and in the hinterland of Eurasian continent [87° 56′ E, 44° 17′ N; elevation: 461 m]. Soils were sampled from 0 to 15 cm, with 15 replications, from three ecosystem sites (saline desert, abandoned farmland and oasis farmland) which cover a relatively wide range of soil salinity and alkalinity (EC: 0.5–12.0 ds m^{-1}; pH: 7.5–9.8), where SR data was collected with an LI-8100 Automated Soil CO_2 flux System (LI-COR, Lincoln, Nebraska, USA) from diverse soil sites, including typical saline site *Solonchaks* and typical alkaline site *Solonetz*, as classified by FAO 2000 [52]. These two typical saline-alkali sites are both in a heavy alkalinity, with a

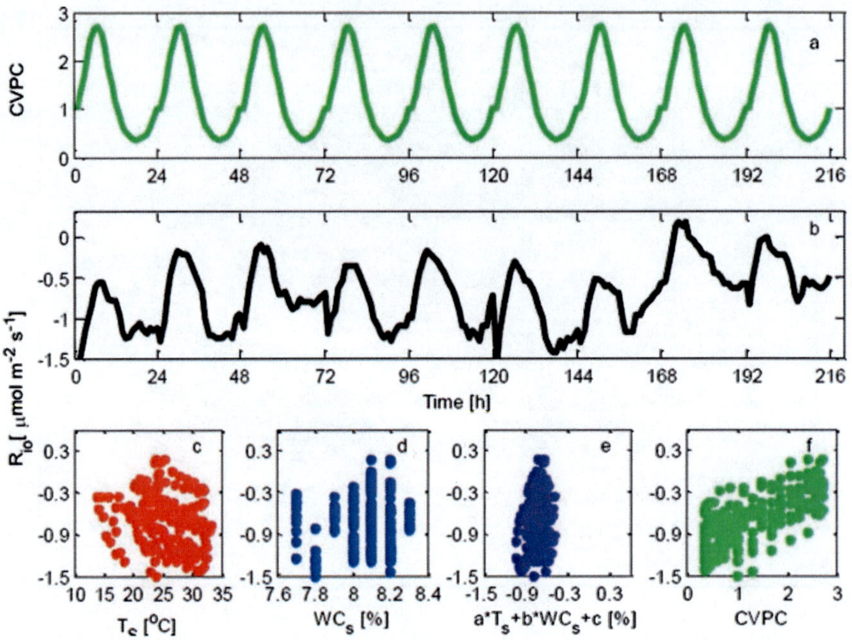

Fig. 3.18 The period character of CVPC (**a**) almost coincides with the period character in the hourly-scale variations of R_{io} (**b**). None of T_S (**c**), WC_S (**d**) and the optimal linear combination of T_S, WC_S (**e**) is better than CVPT (**f**) to describe the temporal pattern of R_{io} in diurnal cycles

notable difference in soil salinity (Solonchaks: pH-9.1, EC: 10.9 ds m^{-1}; Solonetz: pH-10.2, EC: 2.3 ds m^{-1}). Integrated data are shown in Fig. 3.18.

This section has been partly published in [59, 60]. It highlights an unresolved issue beyond reinforcement learning. It was demonstrated that a period character (CVPC) in hourly variations of R_{io} and CVPC almost coincide (Fig. 3.18a, b), and none of TS, WCS and their optimal linear combination is better than CVPC in the description of the temporal pattern of Rio in diurnal cycles (Fig. 3.18c–f). How to utilize the period character to further improve the interpretability of the learning model and reduce bias in reinforcement learning. This should be a next research priority.

References

1. He B, Astolfi L, Valdes-Sosa P A, et al. Electrophysiological Brain Connectivity: Theory and Implementation[J]. IEEE Transactions on Biomedical Engineering, 2019, PP(99):1–1.
2. Fang Y, Chen M, Zheng X. Extracting features from phase space of EEG signals in brain–computer interfaces[J]. Neurocomputing, 2015, 151(10): 1477–1485.
3. Shi M H, Zhou C L, Xie J, et al. Electroencephalogram-based brain-computer interface for the Chinese spelling system: a survey*[J]. Frontiers of Information Technology & Electronic Engineering, 2018, 19(3).

4. Yan Bo, He Shaobo, Sun Kehui. Design of a network Permutation entropy and its applications for chaotic time series and EEG signals [J]. Entropy, 2019, 21, 849; https://doi.org/10.3390/e21090849.
5. Dai Y, Zhang X, Chen Z, et al. Classification of electroencephalogram signals using wavelet-CSP and projection extreme learning machine[J]. Review of Scientific Instruments, 2018, 89(7): 074302.
6. He Shaobo, Sun Kehui, Wang Rixing. Fractional fuzzy entropy algorithm and the complexity analysis for nonlinear time series [J]. Eur. Phys. J. Special Topics, 2018, 227, 943–957.
7. Herman P, Prasad G, McGinnity T M, et al. Comparative analysis of spectral approaches to feature extraction for EEG-based motor imagery classification[J]. IEEE Transactions on Neural Systems and Rehabilitation Engineering, 2008, 16(4): 317–326.
8. Gu Y, Cleeren E, Dan J. Comparison between Scalp EEG and Behind-the-Ear EEG for Development of a Wearable Seizure Detection System for Patients with Focal Epilepsy.[J]. 2018, 18(1):29.
9. J.W. Ahn, Y. Ku, D.Y. Kim, et al. Wearable in-the-ear EEG system for SSVEP-based brain–computer interface[J]. Electronics Letters, 2018, 54(7):413–414.
10. Reyes-Ortiz, J. L., Oneto, L., & Anguita, D. (2015). Big data analytics in the cloud: spark on hadoop vs mpi/openmp on beowulf. Procedia Computer Science, 53, 121–130.
11. Jahromi M G, Parsaei H, Zamani A, et al. Cross Comparison of Motor Unit Potential Features used in EMG Signal Decomposition[J]. IEEE Transactions on Neural Systems and Rehabilitation Engineering, 2018:1–1.
12. Parhi, K. K., & Ayinala, M. (2014). Low-complexity welch power spectral density computation. IEEE Transactions on Circuits and Systems I: Regular Papers, 61(1), 172–182.
13. Harris, F. J. (1978). On the use of windows for harmonic analysis with the discrete fourier transform. Proceedings of the IEEE, 66(1), 51–83.
14. Ayinala, M., Brown, M., & Parhi, K. K. (2012). Pipelined parallel fft architectures via folding transformation. IEEE Transactions on Very Large Scale Integration (VLSI) Systems, 20(6), 1068–1081.
15. Cho, T., & Lee, H. (2013). A High-Speed Low-Complexity Modified, ${\rm Radix-2^{5$ FFT Processor for High Rate WPAN Applications. IEEE International Symposium on Circuits & Systems. IEEE.
16. Wang, Z., Yeo, L. G., Li, W., Yan, Y., Ting, Y., & Tomisawa, M. (2006). A Novel FFT Processor for OFDM UWB Systems. Apccas IEEE Asia Pacific Conference on Circuits & Systems. IEEE.
17. Qiao, S., Hei, Y., Wu, B., & Zhou, Y. (2007). An area and power efficient FFT processor for UWB systems. Wireless Communications, Networking and Mobile Computing, 2007. WiCom 2007. International Conference on. IEEE.
18. Tang, S. N., Tsai, J. W., & Chang, T. Y. (2010). A 2.4-gs/s fft processor for ofdm-based wpan applications. IEEE Transactions on Circuits and Systems II: Express Briefs, 57(6), 0–455.
19. Garrido, M. (2009). A pipelined fft architecture for real-valued signals. IEEE Transactions on Circuits & Systems I Regular Papers, 56(12), 2634–2643.
20. Ayinala, Manohar, Parhi, & Keshab, K. (2013). Fft architectures for real-valued signals based on radix-2(3) and;radix-2(4) algorithms. IEEE Transactions on Circuits & Systems I Regular Papers, 60(9), 2422–2430.
21. Xiao-Ping, L., Zhu-Lin, A. N., & Li-Ping, Z. (2004). Master-slave parallel genetic algorithm framework on mpi. Acta Simulata Systematica Sinica, 16(9), 1938–1827.
22. EEG Data Set. [Online].Available: http://kdd.ics.uci.edu/databases/eeg/eeg.html.
23. Anderson, M.J., Gribble, N.A., 1998. Partitioning the variation among spatial, temporal and environmental components in a multivariate data set. Australian Journal of Ecology 23, 158–67.
24. Baker, J. M., Ochsner, T. E., Venterea, R. T., Griffis, T. J., 2007. Tillage and soil carbon sequestration—what do we really know? Agric Ecosyst Environ 118, 1–5.
25. Baldocchi, D.D., Falge, E., Gu, L., Olson, R., Hollinger, D., Running, D., Anthoni, P.M., Bernhofer, C., Davis, K.J., Evans, R., Fuentes, J.D., Goldstein, A.H., Katul, G.G., Law, B.E., Lee, Z., Malhi, Y., Meyers, T.P., Munger, W., Oechel, W., Paw, U K.T., Pilegaard, K., Schmid, H.P., Valentini, R., Verma, S.B., Vesala, T., Wilson, K.B., Wofsy, S.C., 2001. FLUXNET: a

new tool to study the temporal and spatial variability of ecosystem-scale carbon dioxide, water vapor, and energy flux densities. Bull. Am. Meteorol. Soc. 82, 2415–2434.
26. Ball, B.A., Virginia, R.A., Barrett, J.E., Parsons, A.N., Wall, D.H., 2009. Interactions between physical and biotic factors influence CO2 flux in Antarctic dry valley soils. Soil Biology & Biochemistry 41, 1510–1517.
27. Chapin, F.S., Woodwell, G.M., Randerson, J.D., Rastetter, E.B., Lovett, G.M., Baldocchi, D.D., Clark, D.A., Harmon, M.E., Schimel, D.S., Valentini, R., Wirth, C., Aber, D.J., Cole, J.J., Goulden, M.L., Harden, J.W., Heimann, M., Howarth, R.W., Matson, P.A., McGuire, A.D., Melillo, J.M., Mooney, H.A., Neff, J.C., Houghton, R.A., Pace, M.L., Ryan, M.G., Running, S.W., Sala, O.E., Schlesinger, W.H., Schulze, E.-D., 2006. Reconciling carbon-cycle concepts. Terminol. Methods Ecosyst. 9, 1041–1050.
28. Dawson, J.J.C., Smith, P., 2007. Carbon losses from soil and its consequences for land-use management. Sci Total Environ 382, 165–190.
29. Detwiler, R.P., Hall C.A.S., 1988. Tropical forests and the global carbon cycle. Science 239:42–47.
30. Emmerich, E.W., 2003. Carbon dioxide fluxes in a semiarid environment with high carbonate soils. Agric. Forest Meteorol. 116, 91–102.
31. de Wit, H.A., Palosuo, T., Hylen, G., Liski, J., 2006. A carbon budget of forest biomass and soils in southeast Norway calculated using a widely applicable method. For Ecol Manage 225, 15–26.
32. Falge, E, Baldocchi, D.D., Tenhunen, J., Aubinet, M., Bakwin, P., Berbigier, P., Bernhofer, C., Burba, G., Clement, R., Kenneth, J.D., Elbers, J., Goldstein, A., Grelle, A., Granier, A., Guðmundsson, J., Hollinger, D., Kowalski, A.S., Katul, G., Law, B.E, Malhi, Y., Meyers, T., Monsons, R.K., Mungert, J.W., Oechel, W., Paw, U K.T., Pilegaard, K., Rannik, U¨., Rebmann, C., Suyker, A.E., Valentini, R., Wilson, A., Wofsy, S.C., 2002. Seasonality of ecosystem respiration and gross primary production as derived from FLUXNET measurements. Agric. Forest Meteorol. 113, 53–74.
33. Fisher, M.J., Rao, I.M., Ayarza, M.A., Lascano, C.E., Sanz, J.I., Thomas, R.J., Vera, R.R., 1994. Carbon storage by introduced deep-rooted grasses in the South American savannas. Nature 371, 236–238.
34. Galang, J.S., Zipper, C.E., Prisley, S.P., Galbraith, J.M., Donovan. P.F., 2007. Evaluating terrestrial carbon sequestration options for Virginia. Environmental Management, 39, 139–150.
35. Gombert, P., 2002. Role of karstic dissolution in global carbon cycle. Global Planet. Change 33, 177–184.
36. Goodale, C.L., Davidson, E.A., 2002. Uncertain sinks in the shrubs. Nature 418, 593–594.
37. Grace, J., Malhi, Y., 2002. Carbon dioxide goes with the flow. Nature 416, 594–595.
38. Hastings, S.J., Oechel, W.C., Muhlia-Melo, A., 2005. Diurnal, seasonal and annual variation in the net ecosystem CO2 exchange of a desert shrub community (Sarcocaulescent) in Baja California, Mexico. Global Change Biol. 11, 1–13, https://doi.org/10.1111/j.1365-2486.2005.00951.x.
39. Inglima, I., Alberti, G., Bertolini, T., Vaccari, F.P., Gioli, B., Miglietta, F., Cotrufo, M.F., Peressotti, A., 2009. Precipitation pulses enhance respiration of Mediterranean ecosystems: the balance between organic and inorganic components of increased soil CO2 efflux. Global Change Biol. 15, 1289–1301, https://doi.org/10.1111/j.1365-2486.2008.01793.x.
40. Jasoni R.L., Smith S.D., Arnone J.A., 2005. Net ecosystem CO2 exchange in Mojave Desert shrublands during the eighth year of exposure to elevated CO2. Glob. Change Biol. 11, 749–756.
41. Keeling R.F., Piper S.C., Heimann M., 1996. Global and hemispheric CO2 sinks deduced from changes in atmospheric O2 concentration. Nature 381:218–221.
42. Kessler, T.J., Harvey, C.F., 2001. Global flux of carbon dioxide into groundwater. Geophys Res Lett 28, 279–282.
43. Koizumi, H., Nakadai, T., Usami, Y., Satoh, M., Shiyomi, M., Oikawa, T. 1991. Effect of carbon dioxide concentration on microbial respiration in soil. Ecol Res 6, 227–232.
44. Kowalski, A.S., Serrano-Ortiz, P., Janssens, I.A., Sánchez -Moral, S., Cuezva, S., Domingo, F., Alados-Arboledas, L., 2008. Can flux tower research neglect geochemical CO2 exchange? Agric. Forest Meteorol. 148 (6–7), 1045–1054.

45. Lal R., 2003. Soil erosion and the global carbon budget. Environment International, 29(4), 437–450.
46. Liu R., Li Y., Wang Q.X., 2011. Variations in water and CO2 fluxes over a saline desert in western China. Hydrol. Process. https://doi.org/10.1002/hyp.8147.
47. Liu, Z., Zhao, J., 2000. Contribution of carbonate rock weathering to the atmospheric CO2 sink. Environ. Geol. 39 (9), 1053–1058.
48. Schlesinger, W.H., Belnap, J., Marion, G., 2009. On carbon sequestration in desert ecosystems. Glob. Change Biol. 15, 1488–1490, https://doi.org/10.1111/j.1365-2486.2008.01763.x.
49. Serrano-Ortiz, P., Roland M., Sánchez -Moral, S., Janssens, I.A., Domingo, F., Goddéris Y., Kowalski, A.S., 2010. Hidden, abiotic CO2 flows and gaseous reservoirs in the terrestrial carbon cycle: Review and perspectives Agric. Forest Meteorol. 150, 321–329.
50. Stone, R., 2008. Have desert researchers discovered a hidden loop in the carbon cycle? Science 320, 1409–1410.
51. Wohlfahrt, G., Fenstermaker, L.F., Arnone, J.A., 2008. Large annual net ecosystem CO2 uptake of a Mojave Desert ecosystem. Glob. Change Biol. 14, 1475–1487, https://doi.org/10.1111/j.1365-2486.2008.01593.x.
52. Xie, J., Li, Y., Zhai, C., Li, C.Z.L., 2009. CO2 absorption by alkaline soils and its implication to the global carbon cycle. Environ. Geol. 56, 953–961, https://doi.org/10.1007/s00254-008r-r1197-0.
53. Valentini, R., Matteucci, G., Dolman, A.J., Schulze, E.-D., Rebmann, C., Moors, E.J., Granier, A., Gross, P., Jensen, N.O., Pilegaard, K., Lindroth, A., Grelle, A., Bernhofer, C., Grünwald, T., Aubinet, M., Ceulemans, R., Kowalski, A.S., Vesala, T., Rannik, Ü., Berbigier, P., Loustau, D., Guðndmonson, J., Thorgeirsson, H., Ibrom, A., Morgenstern, K., Clement, R., Monc rieff, J.B., Montagnani, L., Minerbi, S., Jarvis, P.G., 2000. Respiration as the main determinant of carbon balance in European forests. Nature 404, 861–865.
54. Wang, X.H., Piao, S.L., Ciais, P., Janssens, I.A., Reichstein, M., Peng, S.S., Wang, T., 2010. Are ecological gradients in seasonal Q10 of soil respiration explained by climate or by vegetation seasonality? Soil Biology and Biochemistry, 42, 1728–1734.
55. Wolfsy, S.C., 2001. Where has all the carbon gone? Science 292, 2261–2263.
56. Woodbury, P.B., Smith, J.E., Health, L.S., 2007. Carbon sequestration in the U.S. forest sector from 1990 to 2010. For Ecol Manage 241, 14–27.
57. Wang W F, Chen X, Zheng H, et al. Soil CO_2 Uptake in Deserts and Its Implications to the Groundwater Environment. Water, 2016, 8(9), 379; https://doi.org/10.3390/w8090379.
58. Chen X, Wang W F, Luo G P, et al. Time lag between carbon dioxide influx to and efflux from bare saline-alkali soil detected by the explicit partitioning and reconciling of soil CO_2 flux. Stochastic Environmental Research and Risk Assessment, 2013, 27(3): 737–745.
59. Peng An, Wang W, Chen X et al. Introducing a chaotic component in the control system of soil respiration [J]. Complexity, 2020, Article ID: 5310247.
60. Xiong Q, Zhang X M, Wang W F et al. A Parallel algorithm framework for feature extraction of EEG signals on MPI [J]. Computational and Mathematical Methods in Medicine, 2020, Article ID: 9812019.

Chapter 4
Pattern Analysis and Scene Understanding

Abstract This chapter aims to explain the pattern of machine understanding, utilizing medical test as a practical example. The explanation is based on the semantic information theory. After long arguments between positivism and falsificationism, the verification of universal hypotheses was replaced with the confirmation of uncertain major premises. Unfortunately, Hemple proposed the Raven Paradox. Then, Carnap used the increment of logical probability as the confirmation measure. So far, many confirmation measures have been proposed. Measure F proposed by Kemeny and Oppenheim among them possesses symmetries and asymmetries proposed by Elles and Fitelson, monotonicity proposed by Greco et al., and normalizing property suggested by many researchers. Based on the semantic information theory, a measure b^* similar to F is derived from the medical test. Like the likelihood ratio, measures b^* and F can only indicate the quality of channels or the testing means instead of the quality of probability predictions. Furthermore, it is still not easy to use b^*, F, or another measure to clarify the Raven Paradox. For this reason, measure c^* similar to the correct rate is derived. Measure c^* supports the Nicod Criterion and undermines the Equivalence Condition, and hence, can be used to eliminate the Raven Paradox. An example indicates that measures F and b^* are helpful for diagnosing the infection of Novel Coronavirus, whereas most popular confirmation measures are not. Another example reveals that all popular confirmation measures cannot be used to explain that a black raven can confirm "Ravens are black" more strongly than a piece of chalk. Measures F, b^*, and c^* indicate that the existence of fewer counterexamples is more important than more positive examples' existence, and hence, are compatible with Popper's falsification thought.

4.1 Introduction

All traditional areas of machine intelligence can be summarized as pattern analysis, recognition and scene understanding (computer vision and image understanding). These represent the original concepts of the machine brain. Areas such as techniques for visual search, document and handwriting analysis, medical image analysis, video

W. Wang et al., *Interdisciplinary Evolution of the Machine Brain*,
Research on Intelligent Manufacturing,
https://doi.org/10.1007/978-981-33-4244-6_4

and image sequence analysis, content-based retrieval of image and video, face and gesture recognition and relevant specialized hardware and/or software architectures are also covered. At the current stage—artificial intelligence, a universal judgment is equivalent to a hypothetical judgment or a rule, such as "All ravens are black" is equivalent to "For every x, if x is a raven, then x is black". Both can be used as a major premise for a syllogism. Deductive logic needs major premises; however, some major premises for empirical reasoning must be supported by inductive logic. Logical empiricism affirmed that a universal judgment can be verified finally by sense data. Popper said against logical empiricism that a universal judgment could only be falsified rather than be verified. However, for a universal or hypothetical judgment that is not strict, and is therefore uncertain, such as "Almost all ravens are black", "Ravens are black", or "If a man's Coronavirus test is positive, then he is very possibly infected", we cannot say that one counterexample can falsify it. After long arguments, Popper and most logical empiricists reached the identical conclusion [1, 2] that we may use evidence to confirm universal judgments or major premises that are not strict or uncertain.

In 1945, Hemple [3] proposed the confirmation paradox or the Raven Paradox. According to the Equivalence Condition in the classical logic, "If x is a raven, then x is black" (Rule I) is equivalent to "If x is not black, then x is not a raven" (Rule II). A piece of white chalk supports the Rule II, and hence, also supports the Rule I. However, according to the Nicod criterion [4], a black raven supports the Rule I, a non-black raven undermines the Rule I, and a non-raven thing, such as a black cat or a piece of white chalk, is irrelevant to the Rule I. Hence, there exists a paradox between the Equivalence Condition and the Nicod criterion.

To quantize confirmation, both Carnap [1] and Popper [2] proposed their confirmation measures. However, only Carnap's confirmation measures are famous. So far, researchers have proposed many confirmation measures [1, 5–13]. The induction problem seemly has become the confirmation problem. To screen reasonable confirmation measures, Elles and Fitelson [14] proposed **symmetries and asymmetries** as desirable properties; Crupi et al. [8] and Greco et al. [15] suggested **normalization** (for measures between -1 and 1) as a desirable property; Greco et al. [16] proposed **monotonicity** as a desirable property. We can find that only measures F (proposed by Kemeny and Oppenheim) and Z among popular confirmation measures possess these desirable properties. Measure Z was proposed by Crupi et al. [8] as the normalization of some other confirmation measures. It is also called the certainty factor proposed by Shortliffe and Buchanan [7].

When the author of this paper researched semantic information theory [17], he found that an uncertain prediction could be treated as the combination of a clear prediction and a tautology; the combining proportion of the clear prediction could be used as the degree of belief; the degree of belief optimized with a sampling distribution could be regarded as a confirmation measure. This measure is denoted by b^*; it is similar to measure F and also possesses the above-mentioned desirable properties.

Good confirmation measures should possess not only mathematically desirable properties but also practicabilities. We can use medical tests to check their practicabilities. We use the degree of belief to represent the degree to which we believe a major premise and use the degree of confirmation to denote the degree of belief that is optimized by a sample or some examples. The former is subjective, whereas the latter is objective. A medical test provides the test-positive (or the test-negative) to predict if a person or a specimen is infected (or uninfected). Both the test-positive and the test-negative have degrees of belief and degrees of confirmation. In medical practices, there exists an important issue: if two tests provide different results, which test should we believe? For example, when both Nucleic Acid Test (NAT) and CT (Computed Tomography) are used to diagnose the infection of Novel Coronavirus Disease (COVID-19), if the result of NAT is negative and the result of CT is positive, which should we believe? According to the sensitivity and the specificity [18] of a test and the prior probability of the infection, we can use any confirmation measure to calculate the degrees of confirmation of the test-positive and the test-negative. Using popular confirmation measures, can we provide reasonable degrees of confirmation to help us choose a better result from NAT-negative and CT-positive? Can these degrees of confirmation reflect the probability of the infection?

This chapter will show that only measures that are the functions of the likelihood ratio, such as F and b^*, can help us to diagnose the infection or choose a better result that can be accepted by the medical society. However, measures F and b^* do not reflect the probability of the infection. Furthermore, using F, b^*, or another measure, it is still difficult to eliminate the Raven Paradox.

Recently, the author found that the problem with the Raven Paradox is different from the problem with the medical diagnosis. Measures F and b^* indicate how good the testing means are instead of how good the probability predictions are. To clarify the Raven Paradox, we need a confirmation measure that can indicate how good a probability prediction is. The confirmation measure c^* is hence derived. We call c^* a prediction confirmation measure and call b^* a channel confirmation measure. The distinction between Channels' confirmation and predictions' confirmation is similar to yet different from the distinction between Bayesian confirmation and Likelihoodist confirmation [19]. Measure c^* accords with the Nicod criterion and undermines the Equivalence Condition, and hence can be used to eliminate the Raven Paradox.

The main purposes of this paper are:

- to distinguish channel confirmation measures that are compatible with the likelihood ratio and prediction confirmation measures that can be used to assess probability predictions,
- to use a prediction confirmation measure c^* to eliminate the Raven Paradox, and
- to explain that confirmation and falsification may be compatible.

The confirmation methods in this paper are different from popular methods, since:

- Measures b^* and c^* are derived by the semantic information method [17, 20] and the maximum likelihood criterion rather than defined directly.

- Confirmation and statistical learning mutually support so that the confirmation measures can be used not only to assess major premises but also to make probability predictions.

The main contributions of this paper are:

- It clarifies that we cannot use one confirmation measure for two different tasks: (1) to assess (communication) channels, such as medical tests as testing means, and (2) to assess probability predictions, such as to assess "Ravens are black".
- It provides measure c^* that manifests the Nicod criterion and hence provides a new method to clarify the Raven Paradox.

The rest of this paper is organized as follows. Section 4.2 includes background knowledge. It reviews existing confirmation measures, introduces the related semantic information method, and clarifies some questions about confirmation. Section 4.3 derives new confirmation measures b^* and c^* with the medical test as an example. It also provides many confirmation formulas for major premises with different antecedents and consequents. Section 4.4 includes results. It gives some cases to show the characteristics of new confirmation measures, to compare various confirmation measures by applying them to the diagnosis of COVID-19, and to show how an increased example affects the degrees of confirmation with different confirmation measures. Section 4.5 discusses why we can only eliminate the Raven Paradox by measure c^*. It also discusses some conceptual confusion and explains how new confirmation measures are compatible with Popper's falsification thought. Section 4.6 ends with conclusions.

4.2 Background

4.2.1 Statistical Probability, Logical Probability, Shannon's Channel, and Semantic Channel

First we distinguish logical probability and statistical probability. Logical probability of a hypothesis (or a label) is the probability in which the hypothesis is judged to be true, whereas its statistical probability is the probability in which the hypothesis or the label is selected.

Suppose that ten thousand people go through a door. For everyone denoted by x, entrance guards judge if x is elderly. If two thousand people are judged to be elderly, then the logical probability of the predicate "x is elderly" is 2000/10,000 = 0.2. If the task of entrance guards is to select a label for every person from four labels: "Child", "Youth", "Adult", and "Elderly", there may be one thousand people who are labeled "Elderly". The statistical probability of "Elderly" should be 1000/10,000 = 0.1. Why are not two thousand people are labeled "Elderly"? The reason is that some elderly people are labeled "Adult". A person may make two labels be true, such as a 65 years old person makes both "Adult" and "Elderly" be true. That is why the logical probability of a label is often greater than its statistical probability. An extreme example is that the logical probability of a tautology, such as "x is elderly or not elderly", is 1, whereas its statistical probability is almost 0 in general because a tautology is rarely selected. Statistical probability is normalized (the sum is 1), whereas logical probability is not normalized in general [17]. Therefore, we use two different symbols "P" and "T" to distinguish statistical probability and logical probability.

We now consider the Shannon channel [21] between human ages and labels "Child", "Adult", "Youth", "Middle age", "Elderly", and the like.

Let X be a random variable to denote an age and Y be a random variable to denote a label. X takes a value $x \in$ {ages}; Y takes a value $y \in$ {"Child", "Adult", "Youth", "Middle age", "Elderly",...}. Shannon calls the prior probability distribution $P(X)$ (or $P(x)$) the source, and calls $P(Y)$ the destination. There is a Shannon channel $P(Y|X)$ from X to Y. It is a transition probability matrix:

$$P(Y|X) \Leftrightarrow \begin{bmatrix} P(y_1|x_1) & P(y_1|x_2) & \ldots & P(y_1|x_m) \\ P(y_2|x_1) & P(y_2|x_2) & \ldots & P(y_2|x_m) \\ \ldots & \ldots & \ldots & \ldots \\ P(y_n|x_1) & P(y_n|x_2) & \ldots & P(y_n|x_m) \end{bmatrix} \Leftrightarrow \begin{bmatrix} P(y_j|x) \\ P(y_j|x) \\ \ldots \\ P(y_n|x) \end{bmatrix}, \tag{4.1}$$

where $\Longleftrightarrow$ indicates equivalence. This matrix consists of a group of conditional probabilities $P(y_j|x_i)$ ($j = 0, 1, \ldots, n$; $i = 0, 1, \ldots, m$) or a group of transition probability functions (so called by Shannon [21]), $P(y_j|x)$ ($j = 0, 1, \ldots, n$), where y_j is a constant, and x is a variable.

There is also a semantic channel that consists of a group of truth functions. Let $T(\theta_j|x)$ be the truth function of y_j, where θ_j is a model or a set of model parameters, by which we construct $T(\theta_j|x)$. The θ_j is alse explained as a fuzzy sub-set of the domain of x [17]. For example, y_j = "x is young". Its truth function may be

$$T(\theta_j|x) = \exp\left[-(x-20)^2/25\right], \tag{4.2}$$

where 20 and 25 are model parameters. For y_k= "x is elderly", its truth function may be a logistic function:

$$T(\theta_k|x) = 1/\left[1 + \exp\left[-0.2(x - 65)\right]\right], \tag{4.3}$$

where 0.2 and 65 are model parameters. The two truth functions are shown in Fig. 4.1.

According to Tarski's truth theory [22] and Davidson's truth-conditional semantics [23], a truth function can represent the semantic meaning of a hypothesis. Therefore, we call the matrix, which consists of a group of truth functions, a semantic channel:

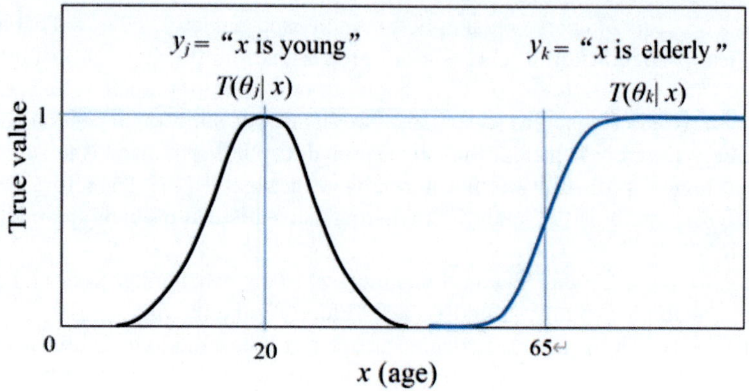

Fig. 4.1 The truth functions of two hypotheses about ages

$$T(\theta|X) \Leftrightarrow \begin{bmatrix} T(\theta_1|x_1) & T(\theta_1|x_2) & \ldots & T(\theta_1|x_m) \\ T(\theta_2|x_1) & T(\theta_2|x_2) & \ldots & T(\theta_2|x_m) \\ \ldots & \ldots & \ldots & \ldots \\ T(\theta_n|x_1) & T(\theta_n|x_2) & \ldots & T(\theta_n|x_m) \end{bmatrix} \Leftrightarrow \begin{bmatrix} T(\theta_1|x) \\ T(\theta_2|x) \\ \ldots \\ T(\theta_n|x) \end{bmatrix}. \quad (4.4)$$

Using a transition probability function $P(y_j|x)$, we can make the probability prediction $P(x|y_j)$ by

$$P(x|y_j) = P(x)P(y_j|x)/P(y_j), \quad (4.5)$$

which is the classical Bayes' formula. Using a truth function $T(\theta_j|x)$, we can also make a probability prediction or produce a likelihood function by

$$P(x|\theta_j) = P(x)T(\theta_j|x)/T(\theta_j), \quad (4.6)$$

where $T(\theta_j)$ is the logical probability of y_j. There is

$$T(\theta_j) = \sum_i P(x_i)T(\theta_j|x_i). \quad (4.7)$$

Equation (4.6) is called the semantic Bayes' formula [17]. The likelihood function is subjective; it may be regarded as the hybird of logical probability and statistical probability.

When the source $P(x)$ is changed, the above formulas for predictions still work. It is easy to prove that $P(x|\theta_j) = P(x|y_j)$ as $T(\theta_j|x) \propto P(y_j|x)$. Since the maximum of $T(\theta_j|x)$ is 1, letting $P(x|\theta_j) = P(x|y_j)$, we can obtain the optimized truth function [17]:

$$T^*(\theta_j|x) = [P(x|y_j)/P(x)]/\max[P(x|y_j)/P(x)] = P(y_j|x)/\max[P(y_j|x)], \tag{4.8}$$

where x is a variable and max(.) is the maximum of the function in brackets (.).

4.2.2 To Review Popular Confirmation Measures

We use h_1 to denote a hypothesis, h_0 to denote its negation, and h to denote one of them. We use e_1 as another hypothesis as the evidence of h_1, e_0 as its negation, and e as one of them. We use $c(e, h)$ to represent a confirmation measure, which means the degree of inductive support. Note that $c(e, h)$ here is used as in [8], where e is on the left, and h is on the right.

In the existing studies of confirmation, logical probability and statistical probability are not definitely distinguished. We still use P for both in introducing popular confirmation measures.

The popular confirmation measures include:

- $D(e_1, h_1) = P(h_1|e_1) - P(h_1)$ [1],
- $M(e_1, h_1) = P(e_1|h_1) - P(e_1)$ [5],
- $R(e_1, h_1) = \log[P(h_1|e_1)/P(h_1)]$ [6],
- $C(e_1, h_1) = P(h_1, e_1) - P(e_1)P(h_1)$ [1],
- $Z(h_1, e_1) = \begin{cases} [P(h_1|e_1) - P(h_1)]/P(h_0), & \text{as } P(h_1|e_1) \geq P(h_1), \\ [P(h_1|e_1) - P(h_1)]/P(h_1), & \text{otherwise}, \end{cases}$ [7, 8],
- $S(e_1, h_1) = P(h_1|e_1) - P(h_1|e_0)$ [9],
- $N(e_1, h_1) = P(e_1|h_1) - P(e_1|h_0)$ [10],
- $L(e_1, h_1) = \log[P(e_1|h_1)/P(e_1|h_0)]$ [11], and
- $F(e_1, h_1) = [P(e_1|h_1) - P(e_1|h_0)]/[P(e_1|h_1) + P(e_1|h_0)]$ [12].

Two measures D and C proposed by Carnap are for incremental confirmation and absolute confirmation respectively. There are more confirmation measures in [8, 24]. Measure F is also denoted by l^* [13], L [8], or k [24]. Most authors explain that probabilities they use, such as $P(h_1)$ and $P(h_1|e_1)$ in D, R, and C, are logical probabilities. Some authors explain that probabilities they use, such as $P(e_1|h_1)$ in F, are statistical probabilities.

Firstly, we need to clarify that confirmation is to assess what kind of evidence supports what kind of hypotheses. Let us have a look at the following three hypotheses:

- Hypothesis 1: $h_1(x)$ = "x is elderly", where x is a variable for an age and $h_1(x)$ is a predicate. An instance $x = 70$ may be the evidence, and the truth value $T(\theta_1|70)$ of proposition $h_1(70)$ should be 1. If $x = 50$, the (uncertain) truth value should be less, such as 0.5. Let e_1 = "$x \geq 60$", true e_1 may also be the evidence that supports h_1 so that $T(\theta_1|e_1) > T(\theta_1)$.

Table 4.1 The numbers of four types of examples for confirmation measures

	e_0	e_1
h_1	b	a
h_0	d	c

- Hypothesis 2: $h_1(x)$ = "If age $x \geq 60$, then x is elderly", which is a hypothetical judgment, a major premise, or a rule. Note that $x = 70$ or $x \geq 60$ is only the evidence of the consequent "x is elderly" instead of the evidence of the rule. The rule's evidence should be a sample with many examples.
- Hypothesis 3: $e_1 \rightarrow h_1$ = "If age $x \geq 60$, then x is elderly", which is the same as Hypothesis 2. The difference is that e_1 = "$x \geq 60$"; h_1 = "x is elderly". The evidence is a sample with many examples like $\{(e_1, h_1), (e_1, h_0), \ldots\}$, or a sampling distribution $P(e, h)$, where P means statistical probability.

Hypothesis 1 has a (uncertain) truth function or a conditional logic probability function between 0 and 1, which is ascertained by our definition or usage. Hypothesis 1 need not be confirmed. Hypothesis 2 or Hypothesis 3 is what we need to confirm. The degree of confirmation is between -1 and 1.

There exist two different understandings about $c(e, h)$:

- Understanding 1: The h is the major premise to be confirmed, and e is the evidence that supports h; h and e are so used by Elles and Fitelson [14].
- Understanding 2: The e and h are those in rule $e \rightarrow h$ as used by Kemeny and Oppenheim [12]. The e is only the evidence that supports consequent h instead of the major premise $e \rightarrow h$ (see Sect. 4.2.3 for further analysis).

Fortunately, although researchers understand $c(e, h)$ in different ways, most researchers agree to use a sample including four types of examples (e_1, h_1), (e_0, h_1), (e_1, h_0), and (e_0, h_0) as the evidence to confirm a rule and to use the four examples' numbers a, b, c, and d (see Table 4.1) to construct confirmation measures. The following statements are based on this common view.

The a is the number of example (e_1, h_1). For example, e_1 = "raven" ("raven" is a label or the abbreviate of "x is a raven") and h_1 = "black"; a is the number of black ravens. Similarly, b is the number of black non-raven things; c is the number of non-black ravens; d is the number of non-black and non-raven things.

To make the confirmation task clearer, we follow Understanding 2 to treat $e \rightarrow h$ = "if e then h" as the rule to be confirmed and replace $c(e, h)$ with $c(e \rightarrow h)$. To research confirmation is to construct or select the function $c(e \rightarrow h) = f(a, b, c, d)$.

To screen reasonable confirmation measures, Elles and Fitelson [14] propose the following symmetries:

- Hypothesis Symmetry (HS): $c(e_1 \rightarrow h_1) = -c(e_1 \rightarrow h_0)$ (two consequents are opposite),
- Evidence Symmetry (ES): $c(e_1 \rightarrow h_1) = -c(e_0 \rightarrow h_1)$ (two antecedents are opposite),
- Commutativity Symmetry (CS): $c(e_1 \rightarrow h_1) = c(h_1 \rightarrow e_1)$, and

- Total Symmetry (TS): $c(e_1 \rightarrow h_1) = c(e_0 \rightarrow h_0)$.

They conclude that only HS is desirable; the other three symmetries are not desirable. We call this conclusion the symmetry/asymmetry requirement. Their conclusion is supported by most researchers. Since TS is the combination of HS and ES, we only need to check HS, ES, and CS. According to this symmetry/asymmetry requirement, only measures L, F, and Z among the measures mentioned above are screened out. It is uncertain whether N can be ruled out by this requirement [15]. See [14, 25, 26] for more discussions about the symmetry/asymmetry requirement.

Greco et al. [15] propose monotonicity as a desirable property. If $f(a, b, c, d)$ does not decrease with a or d and does not increase with b or c, then we say that $f(a, b, c, d)$ has the monotonicity. Measures L, F, and Z have this monotonicity, whereas measures D, M, and N do not have. If we further require that $c(e \rightarrow h)$ are normalizing (between -1 and 1) [8, 12], then only F and Z are screened out. There are also other properties discussed [15, 19]. One is logicality, which means $c(e \rightarrow h) = 1$ without counterexample and $c(e \rightarrow h) = -1$ without positive example. We can also screen out F and Z using the logicality requirement.

Consider the medical test, such as the test for COVID-19. Let e_1 = "positive" (e.g., "x is positive", where x is a specimen), e_0 = "negative", h_1 = "infected" (e.g.,"x is infected"), and h_0 = "uninfected". Then the positive likelihood ratio is $LR^+ = P(e_1|h_1)/P(e_1|h_0)$, which indicates the reliability of the rule $e_1 \rightarrow h_1$. Measures L and F have the one-to-one correspondence with LR:

$$L(e_1 \rightarrow h_1) = \log LR^+; \tag{4.9}$$

$$F(e_1, h_1) = \left(LR^+ - 1\right)/(LR^+ + 1). \tag{4.10}$$

Hence, L and F can also be used to assess the reliability of the medical test. In comparison with LR and L, F can indicate the distance between a test (any F) and the best test ($F = 1$) or the worst test ($F = -1$) better than LR and L. However, LR can be used for the probability predictions of diseases more conveniently [27].

4.2.3 To Distinguish a Major Premise's Evidence and Its Consequent's Evidence

The evidence for the consequent of a syllogism is the minor premise, whereas the evidence for a major premise or a rule is a sample or a sampling distribution $P(e, h)$. In some researchers' studies, e is used sometimes as the minor premise, and sometimes as an example or a sample; h is used sometimes as a consequent, and sometimes as a major premise. Researchers use $c(e, h)$ or $c(h, e)$ instead of $c(e \rightarrow h)$ because they need to avoid the contradiction between the two understandings. However, if we distinguish the two types of evidence, it has no problem to use $c(e \rightarrow h)$. We only need to emphasize that the evidence for a major premise is a sampling distribution $P(e, h)$ instead of e.

If h is used as a major premise and e is used as the evidence (such as in [14, 28]), $-e$ (the negation of e) is puzzling because there are four types of examples instead of two. Suppose $h = p \rightarrow q$ and that e is one of (p, q), $(p, -q)$, $(-p, q)$, and $(-p, q)$. If $(p, -q)$ is the counterexample, and other three examples (p, q), $(-p, q)$ and $(-p, -q)$ are positive examples, which support $p \rightarrow q$, then $(-p, q)$ and $(-p, -q)$ should also support $p \rightarrow -q$ because of the same reason. However, according to HS [14], it is unreasonable that the same evidence supports both $p \rightarrow q$ and $p \rightarrow -q$. In addition, e is a sample with many examples in general. A sample's negation or a sample's probability is also puzzling.

Fortunately, though many researchers say that e is the evidence of a major premise h, they also treat e as the antecedent and treat h as the consequent of a major premise because, only in this way, one can calculate the probabilities or conditional probabilities of e and h for a confirmation measure. Why, then, should we replace $c(e, h)$ with $c(e \rightarrow h)$ to make the task clearer? Section 4.5.3 will show that h used as a major premise will result in the misunderstanding of the symmetry/asymmetry requirement.

4.2.4 Incremental Confirmation or Absolute Confirmation?

Confirmation is often explained as assessing the impact of evidence on hypotheses, or the impact of the premise on the consequent of a rule [14, 19]. However, this paper has a different point of view that confirmation is to assess how well a sample or sampling distribution supports a major premise or a rule; the impact on the rule (e.g., the increment of degree of confirmation) may be made by newly added examples.

Since one can use one or several examples to calculate the degree of confirmation with a confirmation measure, many researchers call their confirmation incremental confirmation [14, 15]. There are also researchers who claim that we need absolute confirmation [29]. This paper supports absolute confirmation.

The problem with incremental confirmation is that the degrees of confirmation calculated are often bigger than 0.5 and are irrelevant to our prior knowledge or a, b, c, and d that we knew before. It is unreasonable to ignore prior knowledge. Suppose that the logical probability of h_1 = "x is elderly" is 0.2; the evidence is one or several people with age(s) $x > 60$; the conditionally logical probability of h_1 is 0.9. With measure D, the degree of confirmation is $0.9 - 0.2 = 0.7$, which is very large and irrelevant to the prior knowledge.

In confirmation function $f(a, b, c, d)$, the numbers a, b, c, and d should be those of all examples including past and current examples. A measure $f(a, b, c, d)$ should be an absolute confirmation measure. Its increment should be

$$\Delta f = f(a + \Delta a,\ b + \Delta b,\ c + \Delta c,\ d + \Delta d) - f(a, b, c, d). \tag{4.11}$$

The increment of the degree of confirmation brought about by a new example is closely related to the number of old examples. Section 4.5.2 will further discuss incremental confirmation and absolute confirmation.

4.2.5 The Semantic Channel and the Degree of Belief of Medical Tests

We now consider the Shannon channel and the semantic channel of the medical test. The relation between h and e is shown in Fig. 4.2.

In Fig. 4.2, h_1 denotes an infected specimen (or person), h_0 denotes an uninfected specimen, e_1 is positive, and e_0 is negative. We can treat e_1 as a prediction "h is infected" and e_0 as a prediction "h is uninfected". In other word, h is a true label or true statement, and e is a prediction or selected label. The x is the observed feature of h; E_1 and E_2 are two sub-sets of the domain of x. If x is in E_1, then e_1 is selected; if x is in E_0, then e_0 is selected.

Figure 4.3 shows the relationship between h and x by two posterior probability distributions $P(x|h_0)$ and $P(x|h_1)$ and the magnitudes of four conditional probabilities (with four colors).

In the medical test, $P(e_1|h_1)$ is called sensitivity [18], and $P(h_0|e_0)$ is called specificity. They ascertain a Shannon channel, which is denoted by $P(e|h)$, as shown in Table 4.2.

We regard predicate $e_1(h)$ as the combination of believable and unbelievable parts (see Fig. 4.4). The truth function of the believable part is $T(E_1|h) \in \{0, 1\}$. The unbelievable part is a tautology, whose truth function is always 1. Then we have the truth functions of predicates $e_1(h)$ and $e_0(h)$:

$$T(\theta_{e1}|h) = b_1' + b_1' T(E_1|h);\ T(\theta_{e0}|h) = b_0' + b_0' T(E_0|h). \tag{4.12}$$

where model parameter b_1' is the proportion of the unbelievable part, and also the truth value for the counter-instance h_0.

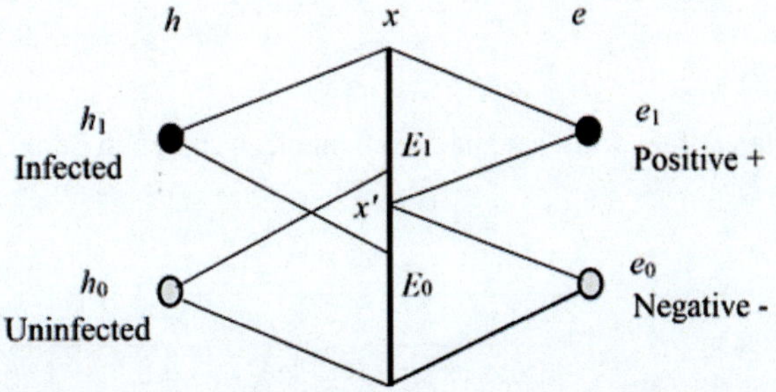

Fig. 4.2 The relationship between positive/negative and infected/uninfected in the medical test

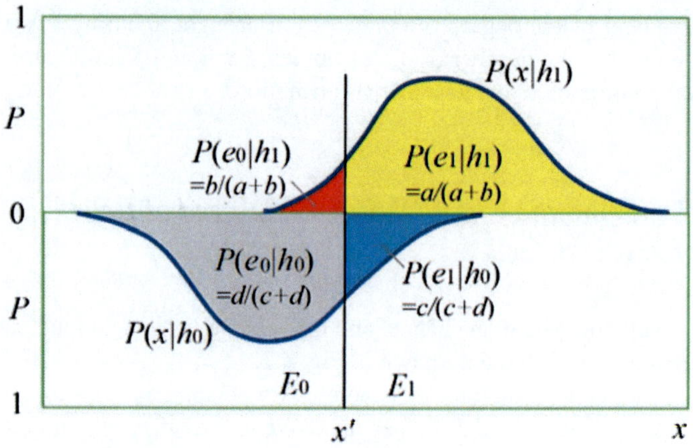

Fig. 4.3 The relationship between two feature distributions and four conditional probabilities for the Shannon channel of the medical test

Table 4.2 Sensitivity and specificity ascertain a Shannon's channel $P(e|h)$

	Negative e_0	Positive e_1
Infected h_1	$P(e_0|h_1) = 1 -$ sensitivity	$P(e_1|h_1) =$ sensitivity
Uninfected h_0	$P(e_0|h_0) =$ specificity	$P(e_1|h_0) = 1 -$ specificity

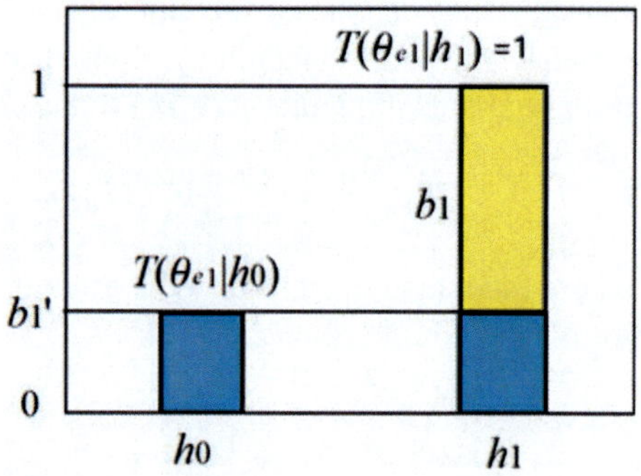

Fig. 4.4 Truth function $T(\theta_{e1}|h)$ includes the believable part with proportion b_1 and the unbelievable part with proportion b_1' ($b_1' = 1 - |b_1|$)

The four truth values form a semantic channel, as shown in Table 4.3 (Tables 4.4, 4.5, 4.6 and 4.7).

Table 4.3 The semantic channel ascertained by b_1' and b_0' for the medical test

	e_0 (Negative)	e_1 (Positive)
h_1 (infected)	$T(\theta e_0|h_1) = b_0'$	$T(\theta_{e1}|h_1) = 1$
h_0 (uninfected)	$T(\theta_{e0}|h_0) = 1$	$T(\theta_{e1}|h_0) = b_1'$

Table 4.4 Predictive probability $P(h_1|\theta_{e1})$ changes with prior probability $P(h_1)$ as $b_1{}^* = 0.9$

	Common people	Risky group	High-risky group
$P(h_1)$	0.001	0.1	0.25
$P(h_1|\theta_{e1})$	0.002	0.19	0.77

Table 4.5 Eight proportions for calculating $b^*(e \to h)$ and $c^*(e \to h)$

	e_0 (Negative)	e_1 (Positive)
h_1 (infected)	$P(e_0|h_1) = b/(a + b)$	$P(e_1|h_1) = a/(a + b)$
h_0 (uninfected)	$P(e_0|h_0) = d/(c + d)$	$P(e_1|h_0) = c/(c + d)$
h_1 (infected)	$P(h_1|e_0) = b/(b + d)$	$P(h_1|e_1) = a/(a + c)$
h_0 (uninfected)	$P(h_0|e_0) = d/(b + d)$	$P(h_0|e_1) = c/(a + c)$

Table 4.6 Channel/prediction confirmation measures expressed by *a*, *b*, *c*, and *d*

	$b^*(e \to h)$ (for Channels, Refer to Fig. 4.3)	$c^*(e \to h)$ (for Predictions, Refer to Fig. 4.7)
$e_1 \to h_1$	$\frac{P(e_1|h_1)-P(e_1|h_0)}{P(e_1|h_1)\vee P(e_1|h_0)}=\frac{ad-bc}{a(c+d)\vee c(a+b)}$	$\frac{P(h_1|e_1)-P(h_0|e_1)}{P(h_1|e_1)\vee P(h_0|e_1)}=\frac{a-c}{a\vee c}$
$e_0 \to h_0$	$\frac{P(e_0|h_0)-P(e_0|h_1)}{P(e_0|h_0)\vee P(e_0|h_1)}=\frac{ad-bc}{d(a+b)\vee b(c+d)}$	$\frac{P(h_0|e_0)-P(h_1|e_0)}{P(h_0|e_0)\vee P(h_1|e_0)}=\frac{d-b}{d\vee b}$

Table 4.7 Converse channel/prediction confirmation measures expressed by *a*, *b*, *c*, and *d*

	$b^*(h \to e)$ (for Converse Channels)	$c^*(h \to e)$ (for Converse Predictions, Refer to Fig. 4.7)
$h_1 \to e_1$	$\frac{P(h_1|e_1)-P(h_1|e_0)}{P(h_1|e_1)\vee P(h_1|e_0)}=\frac{ad-bc}{a(b+d)\vee b(a+c)}$	$\frac{P(e_1|h_1)-P(e_0|h_1)}{P(e_1|h_1)\vee P(e_0|h_1)}=\frac{a-b}{a\vee b}$
$h_0 \to e_0$	$\frac{P(h_0|e_0)-P(h_0|e_1)}{P(h_0|e_0)\vee P(h_0|e_1)}=\frac{ad-bc}{d(a+c)\vee c(b+d)}$	$\frac{P(e_0|h_0)-P(e_1|h_0)}{P(e_0|h_0)\vee P(e_1|h_0)}=\frac{d-c}{d\vee c}$

For medical tests, the logical probability of e_1 is

$$T(\theta_{e1}) = \sum_i P(h_i)T(\theta_{e1}|h_i) = P(h_1) + b_1' P(h_0), \tag{4.13}$$

The likelihood function is

$$P(h|\theta_{e1}) = P(h)T(\theta_{e1}|h)/T(\theta_{e1}). \tag{4.14}$$

$P(h|\theta_j)$ is also the predicted probability of h according to $T(\theta_{e1}|h)$ or the semantic meaning of e_1.

To measure subjective or semantic information, we need subjective probability or logical probability [17]. To measure confirmation, we need statistical probability.

4.2.6 Semantic Information Formulas and the Nicod–Fisher Criterion

According to the semantic information G theory [17], the (amount of) semantic information conveyed by y_j about x_i is defined with the log-normalized-likelihood:

$$I(x_i;\theta_j) = \log\frac{P(x_i|\theta_j)}{P(x_i)} = \log\frac{T(\theta_j|x_i)}{T(\theta_j)}, \tag{4.15}$$

where $T(\theta_j|x_i)$ is the truth value of proposition $y_j(x_i)$ and $T(\theta_j)$ is the logical probability of y_j. If $T(\theta_j|x)$ is always 1, then this semantic information formula becomes Carnap and Bar-Hillel's semantic information formula [30].

In semantic communication, we often see hypotheses or predictions, such as "The temperature is about 10 °C", "The time is about seven o'clock", or "The stock index will go up about 10% next month". Each one of them may be represented by y_j= "x is about x_j." We can express the truth functions of y_j by

$$T(\theta_j|x) = \exp\left[-(x-x_j)^2/(2\sigma^2)\right]. \tag{4.16}$$

Introducing Eq. (4.16) into Eq. (4.15), we have

$$I(x_i;\theta_j) = \log[1/T(\theta_j)] - (x_i - x_j)^2/(2\sigma^2), \tag{4.17}$$

by which we can explain that this semantic information is equal to the Carnap–Bar-Hillel's semantic information minus the squared relative deviation. This formula is illustrated in Fig. 4.5.

Figure 4.5 indicates that the smaller the logical probability is, the more information there is; and the larger the deviation is, the less information there is. Thus, a wrong

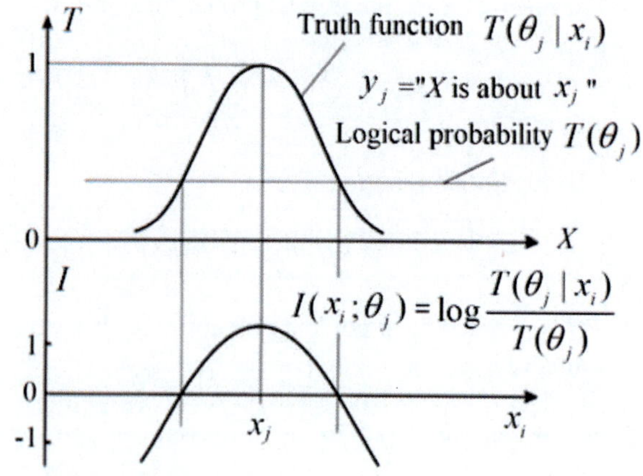

Fig. 4.5 The semantic information conveyed by y_j about x_i

hypothesis will convey negative information. These conclusions accord with Popper's thought (see [2], p. 294).

To average $I(x_i; \theta_j)$, we have generalized Kullback–Leibler information or relative cross-entropy:

$$I(X;\theta_j) = \sum_i P(x_i|y_j) \log \frac{P(x_i|\theta_j)}{P(x_i)} = \sum_i P(x_i|y_j) \log \frac{T(\theta_j|x_i)}{T(\theta_j)}, \tag{4.18}$$

where $P(x|y_j)$ is the sampling distribution, and $P(x|\theta_j)$ is the likelihood function. If $P(x|\theta_j)$ is equal to $P(x|y_j)$, then $I(X; \theta_j)$ reaches its maximum and becomes the relative entropy or the Kullback–Leibler divergence.

Consider medical tests, the semantic information conveyed by e_1 about h becomes

$$I(h_i;\theta_{e1}) = \log \frac{P(h_i|\theta_{e1})}{P(h_i)} = \log \frac{T(\theta_{e1}|h)}{T(\theta_{e1})}. \tag{4.19}$$

The average semantic information is:

$$I(h;\theta_{e1}) = \sum_{i=0}^{1} P(h_i|e_1) \log \frac{P(h_i|\theta_{e1})}{P(h_i)} = \sum_{i=0}^{1} P(h_i|e_1) \log \frac{T(\theta_{e1}|h_i)}{T(\theta_{e1})} \tag{4.20}$$

where $P(h_i|e_1)$ is the conditional probability from a sample.

We now consider the relationship between the likelihood and the average semantic information.

Let **D** be a sample $\{(h(t), e(t))|t = 1 \text{ to } N;\ h(t) \in \{h_0, h_1\};\ e(t) \in \{e_0, e_1\}\}$, which includes two sub-samples or conditional samples $\mathbf{H}_0$ with label e_0 and $\mathbf{H}_1$ with label e_1. When N data points in **D** come from Independent and Identically Distributed random variables, we have the log-likelihood

$$L(\theta_{e1}) = \log P(\mathbf{H}_1|\theta_{e1}) = \log P(h(1), h(2), \ldots, h(N)|\theta_{e1}) = \log \prod_{i=0}^{1} P(h_i|\theta_{e1})^{N_{1i}}$$

$$= N_1 \sum_{i=0}^{1} P(h_i|e_1) \log P(h_i|\theta_{ej}) = -N_1 H(h|\theta_{e1}). \tag{4.21}$$

where N_{1i} is the number of example (h_i, e_1) in **D**; N_1 is the size of $\mathbf{H}_1$. $H(h|\theta_{e1})$ is the cross-entropy. If $P(h|\theta_{e1}) = P(h|e_1)$, then the cross-entropy becomes the Shannon entropy. Meanwhile, the cross-entropy reaches its minimum, and the likelihood reaches its maximum.

Comparing the above two equations, we have

$$I(h;\theta_{e1}) = L(\theta_{e1})/N_1 - \sum_{i=0}^{1} P(h_i|e_1) \log P(h_i) \tag{4.22}$$

which indicates the relationship between the average semantic information and the likelihood. Since the second term on the right side is constant, the maximum likelihood criterion is equivalent to the maximum average semantic information criterion. It is easy to find that a positive example (e_1, h_1) increases the average log-likelihood $L(\theta_{e1})/N_1$; a counterexample (e_1, h_0) decreases it; examples (e_0, h_0) and (e_0, h_1) with e_0 are irrelevant to it.

The Nicod criterion about confirmation is that a positive example (e_1, h_1) supports rule $e_1 \to h_1$; a counterexample (e_1, h_0) undermines $e_1 \to h_1$. No reference exactly indicates if Nicod affirmed that (e_0, h_1) and (e_0, h_1) are irrelevant to $e_1 \to h_1$. If Nicod did not affirm, we can add this affirmation to the criterion, then call the corresponding criterion the Nicod–Fisher criterion, since Fisher proposed the maximum likelihood estimation. From now on, we use the Nicod–Fisher criterion to replace the Nicod criterion.

4.2.7 Selecting Hypotheses and Confirming Rules: Two Tasks from the View of Statistical Learning

Researchers have noted the similarity between most confirmation measures and information measures. One explanation [31] is that information is the average of confirmatory impact. However, this paper gives a different explanation as follows.

There are three tasks in statistical learning: label learning, classification, and reliability analysis. There are similar tasks in inductive reasoning:

- Induction. It is similar to label learning. For uncertain hypotheses, label learning is to train a likelihood function $P(x|\theta_j)$ or a truth function $T(\theta_j|x)$ by a sampling distribution [17]. The Logistic function often used for binary classifications may be treated as a truth function.
- Hypothesis selection. It is like classification according to different criteria.
- Confirmation. It is similar to reliability analysis. The classical methods are to provide likelihood ratios and correct rates (including false rates, as those in Table 4.8) (Tables 4.9, 4.10, 4.11 and 4.12).

Classification and reliability analysis are two different tasks. Similarly, hypothesis selection and confirmation are two different tasks.

In statistical learning, classification depends on the criterion. The often-used criteria are the maximum posterior probability criterion (which is equivalent to the maximum correctness criterion) and the maximum likelihood criterion (which is equivalent to the maximum semantic information criterion [17]). The classifier for binary classifications is

$$e(x) = \begin{cases} e_1, & \text{if } P(\theta_1|x) \geq P(\theta_0|x),\ P(x|\theta_1) \geq P(x|\theta_0),\ \text{or } I(x;\theta_1) \geq I(x;\theta_0); \\ e_0, & \text{otherwise.} \end{cases} \tag{4.23}$$

Table 4.8 PCMs (Prediction Confirmation Measures) are related to different correct rates and false rates in the medical test [18]

PCM	Correct rate positively related to c^*	False rate negatively related to c^*
$c^*(e_1 \to h_1)$	$P(h_1 \mid e_1)$: PPV (Positive Predictive Value)	$P(h_0 \mid e_1)$: FDR (False Discovery Rate)
$c^*(e_0 \to h_0)$	$P(h_0 \mid e_0)$: NPV (Negative Predictive Value)	$P(h_1 \mid e_0)$: FOR (False Omission Rate)
$c^*(h_1 \to e_1)$	$P(e_1 \mid h_1)$: Sensitivity or TPR (True Positive Rate)	$P(e_0 \mid h_1)$: FNR (False Negative Rate)
$c^*(h_0 \to e_0)$	$P(e_0 \mid h_0)$: Specificity or TNR (True Negative Rate)	$P(e_1 \mid h_0)$: FPR (False Positive Rate)

Table 4.9 Three examples to show the differences between different confirmation measures

Ex.	a, b, c, d	D	M	R	C	Z	S	N	L	F	b^*	c^*
1	20, 180, 8, 792	0.514	0.072	1.84	0.014	0.643	0.529	0.09	**3.32**	**0.818**	**0.9**	0.8
2	200, 0, 720, 80	0.017	0.08	0.12	0.016	0.022	0.217	0.1	**0.152**	**0.053**	**0.1**	−0.722
3	10, 0, 90, 900	0.09	0.9	3.32	0.009	**0.091**	0.1	0.091	3.46	**0.833**	**0.91**	−0.9

Table 4.10 Sensitivities and specificities of NAT (Nucleic Acid Test) and CT for COVID-19

	Sensitivity	Specificity
NAT	0.5	0.95
CT	0.8	0.75

Table 4.11 Improved diagnosis (for final positive or negative) according to NAT and CT

	NAT-negative, $b_0^* = 0.47$	NAT-positive, $b_1^* = 0.9$
CT-positive, $b_1^*=0.69$	Final positive (changed)	Final positive
CT-negative, $b_0^*=0.73$	Final negative	Final positive

Table 4.12 Various confirmation measures for assessing the results of NAT and CT

	D	M	Z	S	C	N	F	b^*	c^*
c(NAT−)	0.10	0.11	0.40	0.62	0.08	0.45	0.31	0.47	0.83
c(NAT+)	0.52	0.34	0.69	0.62	0.08	0.45	0.82	0.90	0.70
c(CT−)	0.17	0.14	0.67	0.43	0.10	0.55	0.58	0.73	0.91
c(CT+)	0.27	0.41	0.36	0.43	0.10	0.55	0.52	0.69	0.06
c(CT+)>c(NAT−)			No	No					No
c(NAT+)>c(CT−)					No	No			No

After the above classification, we may use information criterion to assess how well e_j is used to predict h_j:

$$I^*(h_j; \theta_{ej}) = I(h_j; e_j) = \log \frac{P(h_j|e_j)}{P(h_j)} = \log \frac{P(e_j|h_j)}{P(e_j)}$$
$$= \log P(h_j|e_j) - \log P(h_j) = \log P(e_j|h_j) - \log P(e_j)$$
$$= \log P(h_j, e_j) - \log[P(h_j)P(e_j)], \tag{4.24}$$

where I^* means optimized semantic information. With information amounts $I(h_i; \theta_{ej})$ $(i, j = 0, 1)$, we can optimize the classifier [17]:

$$e_j^* = f(x) = \arg\max_{e_j}[P(h_0|x)I(h_0; \theta_{ej}) + P(h_1|x)I(h_1; \theta_{ej})]. \tag{4.25}$$

The new classifier will provide the new Shannon's channel $P(e|h)$. The maximum mutual information classification can be achieved by repeating Eqs. (4.23) and (4.25) [17, 32].

With the above classifiers, we can make prediction e_j= "x is h_j" according to x. To tell information receivers how reliable the rule $e_j \rightarrow h_j$ is, we need the likelihood ratio *LR* to indicate how good the channel is or need the correct rate to indicate how good the probability prediction is. Confirmation is similar. We need to provide a confirmation measure similar to *LR*, such as *F*, and a confirmation measure similar to the correct rate. The difference is that the confirmation measures should change between −1 and 1.

According to above analyses, it is easy to find that confirmation measures *D*, *N*, *R*, and *C* are more like information measures for assessing and selecting predictions instead of confirming rules. *Z* is their normalization [8]; it seems between an information measure and a confirmation measure. However, confirming rules is different from measuring predictions' information; it needs the proportions of positive examples and counterexamples.

4.3 Two Novel Confirmation Measures

4.3.1 To Derive Channel Confirmation Measure b*

We use the maximum semantic information criterion, which is consistent with the maximum likelihood criterion, to derive the channel confirmation measure. According to Eqs. (4.13) and (4.18), the average semantic information conveyed by e_1 about h is

$$I(h;\theta_{e1}) = P(h_0|e_1)\log\frac{b_1'}{P(h_1 + b_1'P(h_0)} + P(h_1|e_1)\log\frac{1}{P(h_1 + b_1'P(h_0)} \tag{4.26}$$

Letting $dI(h;\theta_{e1})/db_1' = 0$, we can obtain the optimized b_1':

$$b_1'^{*} = \frac{P(h_0|e_1)}{P(h_0)} \bigg/ \frac{P(h_1|e_1)}{P(h_1)}, \tag{4.27}$$

where $P(h_1|e_1)/P(h_1) \geq P(h_0|e_1)/P(h_0)$. The b'^{*} can be called a disconfirmation measure. Letting both the numerator and the denominator multiply by $P(e_1)$, the above formula becomes:

$$b_1'^{*} = P(e_1|h_0)/P(e_1|h_1) = (1 - \text{specificity})/\text{sensibility} = 1/LR^{+}. \tag{4.28}$$

According to the semantic information G theory [17], when a truth function is proportional to the corresponding transition probability function, e.g., $T^{*}(\theta_{e1}|h)$ $P(e_1|h)$, the average semantic information reaches its maximum. Using $T^{*}(\theta_{e1}|h)$ $P(e_1|h)$, we can directly obtain

$$\frac{b_1'^{*}}{P(e_1|h_0)} = \frac{1}{P(e_1|h_1)} \tag{4.29}$$

and Eq. (4.28). We call

$$b_1^{*} = 1 - b_1'^{*} = [P(e_1|h_1) - P(e_1|h_0)]/P(e_1|h_1) \tag{4.30}$$

the degree of confirmation of the rule $e_1 \to h_1$. Considering $P(e_1|h_1) < P(e_1|h_0)$, we have

$$b_1^{*} = b_1'^{*} - 1 = [P(e_1|h_0) - P(e_1|h_1)]/P(e_1|h_0). \tag{4.31}$$

Combining the above two formulas, we obtain

$$b_1^{*} = b^{*}(e_1 \to h_1) = \frac{P(e_1|h_1) - P(e_1|h_0)}{\max[P(e_1|h_1), P(e_1|h_0)]} = \frac{LR^{+} - 1}{\max(LR^{+}, 1)}. \tag{4.32}$$

Since

$$b_1^{*} = b^{*}(e_1 \to h_0) = \frac{P(e_1|h_0) - P(e_1|h_1)}{\max[P(e_1|h_0), P(e_1|h_1)]} = -b^{*}(e_1 \to h_1), \tag{4.33}$$

the $b_1{}^{*}$ possesses HS or Consequent Symmetry.

In the same way, we obtain

$$b_0^* = b^*(e_0 \to h_0) = \frac{P(e_0|h_0) - P(e_0|h_1)}{\max[P(e_0|h_0), P(e_0|h_1)]} = \frac{LR^- - 1}{\max(LR^-, 1)}. \quad (4.34)$$

Using Consequent Symmetry, we can obtain $b^*(e_1 \to h_0) = -b^*(e_1 \to h_1)$ and $b^*(e_0 \to h_1) = -b^*(e_0 \to h_0)$.

Using measure b^* or F, we can answer the question: if the result of NAT is negative and the result of CT is positive, which should we believe? Section 4.4.2 will provide the answer that is consistent with the improved diagnosis of COVID-19 in Wuhan.

Compared with F, b^* is better for probability predictions. For example, from b_1^* > 0 and $P(h)$, we obtain

$$P(h_1|\theta_{e1}) = P(h_1)/\left[P(h_1) + b_1'^* P(h_0)\right] = P(h_1)/\left[1 - b_1'^* P(h_0)\right]. \quad (4.35)$$

This formula is much simpler than the classical Bayes' formula (see Eq. (4.5)).

If $b_1^* = 0$, then $P(h_1|\theta_{e1}) = P(h_1)$. If $b_1^* < 0$, then we can make use of HS or Consequent Symmetry to obtain $b_{10}^* = b_1^*(e_1 \to h_0) = |b_1^*(e_1 \to h_1)| = |b_1^*|$. Then we have

$$P(h_0|\theta_{e1}) = P(h_0)/\left[P(h_0) + b_{10}'^* P(h_1)\right] = P(h_0)/\left[1 - b_{10}'^* P(h_1)\right]. \quad (4.36)$$

We can also obtain $b_1^* = 2F_1/(1 + F_1)$ from $F_1 = F(e_1 \to h_1)$ for the probability prediction $P(h_1|\theta_{e1})$, but the calculation of probability predictions with F_1 is a little complicated.

So far, it is still problematic to use b^*, F, or another measure to handle the Raven Paradox. For example, as shown in Table 4.13, the increment of $F(e_1 \to h_1)$ caused by $\Delta d = 1$ is 0.348 − 0.333, whereas the increment caused by $\Delta a = 1$ is 0.340 − 0.333. The former is greater than the latter, which means that a piece of white chalk can support "Ravens are black" better than a black raven. Hence measure F does not accord with the Nicod–Fisher criterion. Measures b^* and Z do not either.

Why does not measure b^* and F accord with the Nicod–Fisher criterion? The reason is that the likelihood $L(\theta_{e1})$ is related to prior probability $P(h)$, whereas b^* and F are irrelevant to $P(h)$.

Table 4.13 How confirmation measures are affected by $\Delta a = 1$ and $\Delta d = 1$

	$f(a, b, c, d)$	$a = d = 20$ $b = c = 10$	$\Delta a = 1$ $\Delta d = 0$	$\Delta d = 1$ $\Delta a = 0$	$\Delta f/\Delta a - \Delta f/\Delta d$
$D(e_1 \to h_1)$	$a/(a+c)-(a+b)/n$	0.167	0.169	0.175	−0.006
$M(e_1 \to h_1)$	$a/(a+b)-(a+c)/n$	0.167	0.169	0.175	−0.006
$C(e_1 \to h_1)$	$a/n - (a+c)(a+b)/n^2$	0.083	0.086	0.086	0
$Z(e_1 \to h_1)$	$D(e_1 \to h_1)/[(c+d)/n]$	0.333	0.344	0.344	0
$S(e_1 \to h_1)$	$a/(a+c)-b(b+d)$	0.333	0.334	0.344	0
$N(e_1 \to h_1)$	$a/(a+b) - c/(c+d)$	0.333	0.334	0.344	0
$F(e_1 \to h_1)$	$(ad-bc)/(ad+bc+2ac)$	0.333	0.340	0.348	−0.007
LR^+	$[a/(a+b)]/[c/(c+d)]$	2	2.03	2.07	−0.034
$c^*(e_1 \to h_1)$	$(a-c)/\max(a, c)$	**0.5**	**0.524**	**0.5**	**0.024 > 0**

4.3.2 To Derive Prediction Confirmation Measure *c**

Statistics not only uses the likelihood ratio to indicate how reliable a testing means (as a channel) is but also uses the correct rate to indicate how reliable a probability prediction is. Measure F and b^* like LR cannot indicate the quality of a probability prediction. Most other measures have similar problems.

For example, we assume that an NAT for COVID-19 [33] has sensitivity $P(e_1|h_1) = 0.5$ and specificity $P(e_0|h_0) = 0.95$. We can calculate $b_1{'}^* = 0.1$ and $b_1^* = 0.9$. When the prior probability $P(h_1)$ of the infection changes, predicted probability $P(h_1|\theta_{e1})$ (see Eq. (4.35)) changes with the prior probability, as shown in Table 4.4. We can obtain the same results using the classical Bayes' formula (see Eq. (4.5)).

Data in Table 4.4 show that measure b^* cannot indicate the quality of probability predictions. Therefore, we need to use $P(h)$ to construct a confirmation measure that can reflect the correct rate.

We now treat probability prediction $P(h|\theta_{e1})$ as the combination of a believable part with proportion c_1 and an unbelievable part with proportion $c_1{'}$, as shown in Fig. 4.6. We call c_1 the degree of belief of the rule $e_1 \rightarrow h_1$ as a prediction.

When the prediction accords with the fact, e.g., $P(h|\theta_{e1}) = P(h|e_1)$, c_1 becomes c_1^*. The degree of disconfirmation for predictions is

$$\begin{aligned} c_1^*(e_1 \rightarrow h_1) &= P(h_0|e_1)/P(h_1|e_1), \text{ if } P(h_0|e_1) \le P(h_1|e_1); \\ c_1^*(e_1 \rightarrow h_1) &= P(h_1|e_1)/P(h_0|e_1), \text{ if } P(h_1|e_1) \le P(h_0|e_1). \end{aligned} \tag{4.37}$$

Further, we have the prediction confirmation measure

$$\begin{aligned} c_1^* = c^*(e_1 \rightarrow h_1) &= \frac{P(h_1|e_1) - P(h_0|e_1)}{\max(P(h_1|e_1), P(h_0|e_1))} \\ &= \frac{2P(h_1|e_1) - 1}{\max(P(h_1|e_1), 1 - P(h_1|e_1))} = \frac{2CR_1 - 1}{\max(CR_1, 1 - CR_1)}. \end{aligned} \tag{4.38}$$

where $CR_1 = P(h_1|\theta_{e1}) = P(h_1|e_1)$ is the correct rate of rule $e_1 \rightarrow h_1$. This correct rate means that the probability of h_1 we predict as $x \in E_1$ is CR_1. Letting both the

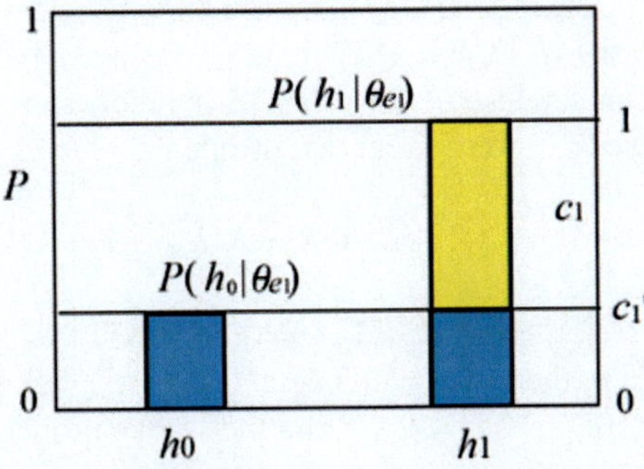

Fig. 4.6 Likelihood function $P(h|\theta_{e1})$ may be regarded as a believable part plus an unbelievable part

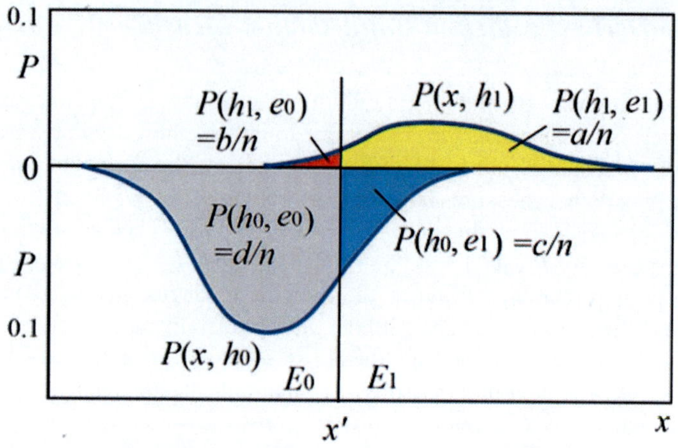

Fig. 4.7 The numbers of positive examples and counterexamples for $c^*(e_0 \to h_0)$ (see the left side) and $c^*(e_1 \to h_1)$ (see the right side)

numerator and denominator of Eq. (4.38) multiply by $P(e_1)$, we obtain

$$c_1^* = c^*(e_1 \to h_1) = \frac{P(h_1, e_1) - P(h_0, e_1)}{\max(P(h_1, e_1), P(h_0, e_1))} = \frac{a-c}{\max(a, c)}. \quad (4.39)$$

The sizes of four areas covered by two curves in Fig. 4.7 may represent *a*, *b*, *c*, and *d*.

In like manner, we obtain

$$c_0^* = c^*(e_0 \to h_0) = \frac{P(h_0, e_0) - P(h_1, e_0)}{\max(P(h_0, e_0), P(h_1, e_0))} = \frac{d-b}{\max(d, b)}. \quad (4.40)$$

Making use of Consequent Symmetry, we can obtain $c^*(e_1 \to h_0) = -\, c^*(e_1 \to h_1)$ and $c^*(e_0 \to h_1) = -\, c^*(e_0 \to h_0)$.

In Fig. 4.7, the sizes of the two areas covered by two curves are $P(h_0)$ and $P(h_1)$, which are different. If $P(h_0) = P(h_1) = 0.5$, then prediction confirmation measure c^* is equal to channel confirmation measure b^*.

Using measure c^*, we can directly assess the quality of the probability predictions. For $P(h_1|\theta_{e1}) = 0.77$ in Table 4.4, we have $c_1^* = (0.77 - 0.23)/0.77 = 0.701$. We can also use c^* for probability predictions. When $c_1^* > 0$, according to Eq. (4.39), we have the correct rate of rule $e_1 \to h_1$:

$$CR_1 = P(h_1|\theta_{e1}) = 1/(1+c_1^*) = 1/(2 - c_1^*) \quad (4.41)$$

For example, if $c_1^* = 0.701$, then $CR_1 = 1/(2 - 0.701) = 0.77$. If $c^*(e_1 \to h_1) = 0$, then $CR_1 = 0.5$. If $c^*(e_1 \to h_1) < 0$, we may make use of HS to have $c_{10}^* = c^*(e_1 \to h_0) = |c^*_1|$, and then make probability prediction:

$$P(h_0|\theta_{e1}) = 1/(2 - c_{10}^*),$$
$$P(h_1|\theta_{e1}) = 1 - P(h_0|\theta_{e1}) = (1 - c_{10}^*)/(2 - c_{10}^*). \tag{4.42}$$

We may define another prediction confirmation measure by replacing operation max() with +:

$$c_{F1} = c_F^*(e_1 \rightarrow h_1) = \frac{P(h_1|e_1) - P(h_0|e_1)}{P(h_1|e_1) + P(h_0|e_1)} = P(h_1|e_1) - P(h_0|e_1)$$
$$= \frac{P(h_1, e_1) - P(h_0, e_1)}{P(e_1)} = \frac{a - c}{a + c}. \tag{4.43}$$

The c_F* is also convenient for probability predictions when $P(h)$ is certain. There is

$$P(h_1|\theta_{e1}) = CR_1 = (1+c_{F1}^*)/2;$$
$$P(h_0|\theta_{e1}) = 1 - CR_1 = (1 - c_{F1}^*)/2. \tag{4.44}$$

However, when $P(h)$ is variable, we should still use b* with $P(h)$ for probability predictions.

It is easy to prove that $c^*(e_1 \rightarrow h_1)$ and $c_F{}^*(e_1 \rightarrow h_1)$ possess all the above-mentioned desirable properties.

4.3.3 Converse Channel/Prediction Confirmation Measures $b^*(h \rightarrow e)$ and $c^*(h \rightarrow e)$

Greco et al. [19] divide confirmation measures into

- Bayesian confirmation measures with $P(h|e)$ for $e \rightarrow h$,
- Likelihoodist confirmation measures with $P(e|h)$ for $e \rightarrow h$,
- converse Bayesian confirmation measures with $P(h|e)$ for $h \rightarrow e$, and
- converse Likelihoodist confirmation measures with $P(e|h)$ for $h \rightarrow e$.

Similarly, this paper divides confirmation measures into

- channel confirmation measure $b^*(e \rightarrow h)$,
- prediction confirmation measure $c^*(e \rightarrow h)$,
- converse channel confirmation measure $b^*(h \rightarrow e)$, and
- converse prediction confirmation measure $c^*(h \rightarrow e)$.

We now consider $c^*(h_1 \rightarrow e_1)$. The positive examples' proportion and the counterexamples' proportion can be found in the upside of Fig. 4.7. Then we have

$$c^*(h_1 \rightarrow e_1) = \frac{P(e_1|h_1) - P(e_0|h_1)}{\max(P(e_1|h_1), P(e_0|h_1))} = \frac{a - b}{\max(a, b)}. \tag{4.45}$$

The correct rate reflected by $c^*(h_1 \rightarrow e_1)$ is sensitivity or true positive rate $P(h_1|e_1)$. The correct rate reflected by $c^*(h_0 \rightarrow e_0)$ is specificity or true negative rate $P(h_0|e_0)$.

Consider the converse channel confirmation measure $b^*(h_1 \rightarrow e_1)$. Now the source is $P(e)$ instead of $P(h)$. We may swap e_1 with h_1 in $b^*(e_1 \rightarrow h_1)$ or swap a with d and b with c in $f(a, b, c, d)$ to obtain

$$b^*(h_1 \rightarrow e_1) = \frac{P(h_1|e_1) - P(h_1|e_0)}{P(h_1|e_1) \vee P(h_1|e_0)} = \frac{ad - bc}{a(b+d) \vee b(a+c)} \tag{4.46}$$

where $\vee$ is the operator for the maximum of two numbers and is used to replace max(). There are also four types of converse channel/prediction confirmation formulas with a, b, c, and d (see Table 4.7). Due to Consequent Symmetry, there are the eight types of converse channel/prediction confirmation formulas altogether.

4.3.4 Eight Confirmation Formulas for Different Antecedents and Consequents

Table 4.5 shows the positive examples' and counterexamples' proportions needed by measures b^* and c^*.

Table 4.6 provides four types of confirmation formulas with a, b, c, and d for rule $e \rightarrow h$, where function max() is replaced with the operator $\vee$.

These confirmation measures are related to the misreporting rates of the rule $e \rightarrow h$. For example, smaller $b^*(e_1 \rightarrow h_1)$ or $c^*(e_1 \rightarrow h_1)$ means that the test shows positive for more uninfected people.

Table 4.7 includes four types of confirmation measures for $h \rightarrow e$.

These confirmation measures are related to the underreporting rates of the rule $h \rightarrow e$. For example, smaller $b^*(h_1 \rightarrow e_1)$ or $c^*(h_1 \rightarrow e_1)$ means that the test shows negative for more infected people. Underreports are more serious problems.

Each of the eight types of confirmation measures in Tables 4.6 and 4.7 has its consequent-symmetrical form. Therefore, there are 16 types of function $f(a, b, c, d)$ altogether for confirmation.

In a prediction and converse prediction confirmation formula, the conditions of two conditional probabilities are the same; they are the antecedents of rules so that a confirmation measure c^* only depends on the two numbers of positive examples and counterexamples. Therefore, these measures accord with the Nicod–Fisher criterion.

If we change "$\vee$" into "+" in $f(a, b, c, d)$, then measure b^* becomes measure b_F^* $= F$, and measure c^* becomes measure c_F^*. For example,

$$c_F^*(e_1 \rightarrow h_1) = (a - c)/(a + c). \tag{4.47}$$

4.3.5 Relationship Between Measures b* and F

Measure b^* is like measure F. The two measures changes with likelihood ratio *LR*, as shown in Fig. 4.8.

Measure F has four confirmation formulas for different antecedents and consequents [8], which are related to measure $b_F{}^*$ as follows:

$$F(e_1 \to h_1) = \frac{P(e_1|h_1) - P(e_1|h_0)}{P(e_1|h_1) + P(e_1|h_0)} = \frac{ad - bc}{ad + bc + 2ac} = b_F^*(e_1 \to h_1) \quad (4.48)$$

$$F(h_1 \to e_1) = \frac{P(h_1|e_1) - P(h_1|e_0)}{P(h_1|e_1) + P(h_1|e_0)} = \frac{ad - bc}{ad + bc + 2ab} = b_F^*(h_1 \to e_1) \quad (4.49)$$

$$F(e_0 \to h_0) = \frac{P(e_0|h_0) - P(e_0|h_1)}{P(e_0|h_0) + P(e_0|h_1)} = \frac{ad - bc}{ad + bc + 2bd} = b_F^*(e_0 \to h_0) \quad (4.50)$$

$$F(h_0 \to e_0) = \frac{P(h_0|e_0) - P(h_0|e_1)}{P(h_0|e_0) + P(h_0|e_1)} = \frac{ad - bc}{ad + bc + 2cd} = b_F^*(h_0 \to e_0) \quad (4.51)$$

F is equivalent to $b_F{}^*$. Measure b^* has all the above-mentioned desirable properties as well as measure F. The differences are that measure b^* has a greater absolute value than measure F; measure b^* can be used for probability predictions more conveniently (see Eq. (4.35)).

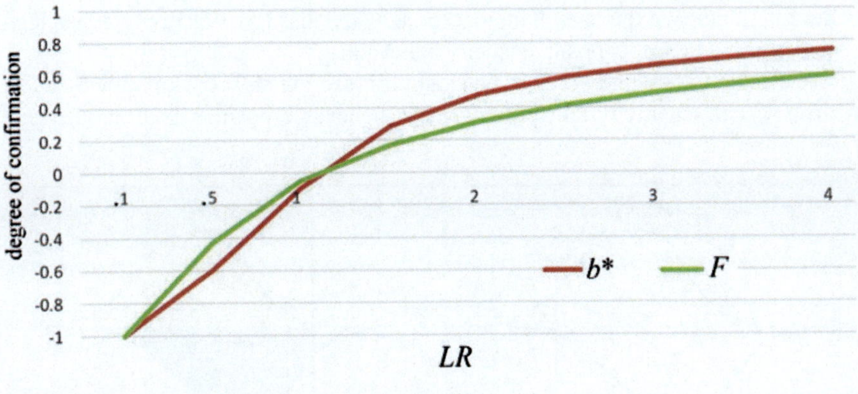

Fig. 4.8 Measures b^* and F change with likelihood ratio *LR*

4.3.6 Relationships Between Prediction Confirmation Measures and Some Medical Test's Indexes

Channel confirmation measures are related to likelihood ratios, whereas Prediction Confirmation Measures (PCMs) including converse PCMs are related to correct rates and false rates in the medical test.

To help us understand the significances of PCMs in the medical test, Table 4.8 shows that each PCM is related to which correct rate and which false rate.

The false rates related to PCMs are the misreporting rates of the rule $e \rightarrow h$, whereas the false rates related to converse PCMs are the underreporting rates of the rule $h \rightarrow e$. For example, False Discovery Rate $P(h_0|e_1)$ is also the misreporting rate of rule $e_1 \rightarrow h_1$; False Negative Rate $P(e_0|h_1)$ is also the underreporting rate of rule $h_1 \rightarrow e_1$.

4.4 Pattern Analysis: A Practical Application

4.4.1 Using Three Examples to Compare Various Confirmation Measures

In China's war against COVID-19, people often ask the question: since the true positive rate, e.g., sensitivity, of NAT is so low (less than 0.5), why do we still believe it? Medical experts explain that though NAT has low sensitivity, it has high specificity, and hence its positive is very believable.

We use the following two extreme examples (see Fig. 4.9) to explain why a test with very low sensitivity can provide more believable positive than another test

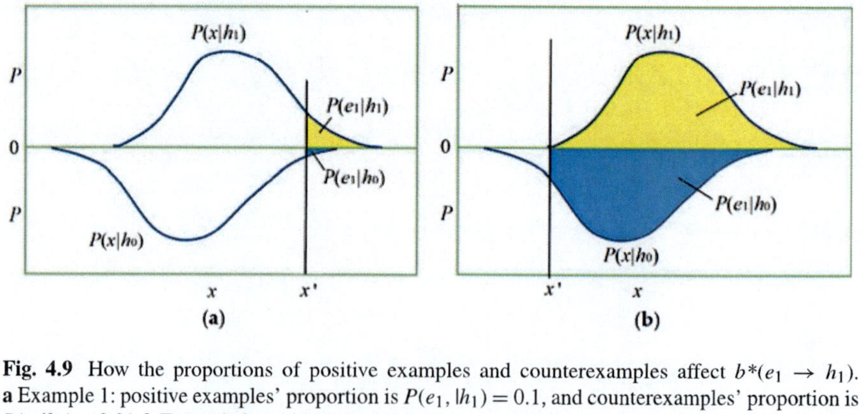

Fig. 4.9 How the proportions of positive examples and counterexamples affect $b^*(e_1 \rightarrow h_1)$. **a** Example 1: positive examples' proportion is $P(e_1, |h_1) = 0.1$, and counterexamples' proportion is $P(e_1|h_0) = 0.01$. **b** Example 2: positive examples' proportion is $P(e_1, |h_1) = 1$, and counterexamples' proportion is $P(e_1|h_0) = 0.9$

with very high sensitivity, and whether popular confirmation measures support this conclusion.

In Example 1, $b^*(e_1 \rightarrow h_1) = (0.1 - 0.01)/0.1 = 0.9$, which is very large. In Example 2, $b^*(e_1 \rightarrow h_1) = (1 - 0.9)/1 = 0.1$, which is very small. The two examples indicate that fewer counterexamples' existence is more important to b^* than more positive examples' existence. Measures F, c^*, and c_F^* also possess this characteristic, which is compatible with the Logicality requirement [15]. However, most confirmation measures do not possess this characteristic.

We supposed $P(h_1) = 0.2$ and $n = 1000$ and then calculated the degrees of confirmation with different confirmation measures for the above two examples, as shown in Table 4.9, where the base of log for R and L is 2. Table 4.9 also includes Example 3 (e.g., Ex. 3), in which $P(h_1)$ is 0.01. Example 3 reveals the difference between Z and b^* (or F).

Data for Examples 1 and 2 show that L, F and b^* give Example 1 a much higher rating than Example 2, whereas M, C, and N give Example 2 a higher rating than Example 1 (see red numbers). The excel file for Tables 4.9, 4.12, and 4.13 can be find in **Supplementary Material**.

In Examples 2 and 3, where $c > a$ (counterexamples are more than positive examples), only the values of $c^*(e_1 \rightarrow h_1)$ are negative. The negative values should be reasonable for assessing probability predictions when counterexamples are more than positive examples.

The data for Example 3 show that when $P(h_0) = 0.99 \gg P(h_1) = 0.01$, measure Z is very different from measures F and b^* (see blue numbers) because F and b^* are independent of $P(h)$ unlike Z.

Although measure L (log-likelihood ratio) is compatible with F and b^*, its values, such as 3.32 and 0.152, are not intuitionistic as well as the values of F or b^*, which are normalizing.

4.4.2 *Using Measures b* to Explain Why and How CT Is also Used to Test COVID-19*

The COVID-19 outbreak in Wuhan of China in 2019 and 2020 has infected many people. In the early stage, only NAT was used to diagnose the infection. Later, many doctors found that NAT often failed to report the viral infection. Because this test has low sensitivity (which may be less than 0.5) and high specificity, we can confirm the infection when NAT is positive, but it is not good for confirming the non-infection when NAT is negative. That means that NAT-negative is not believable. To reduce the underreports of the infection, CT gained more attention because CT had higher sensitivity than NAT.

When both NAT and CT were used in Wuhan, doctors improved the diagnosis, as shown in Fig. 4.10 and Table 4.11. If we diagnose the infection according to confirmation measure b^*, will the diagnosis be the same as the improved diagnosis?

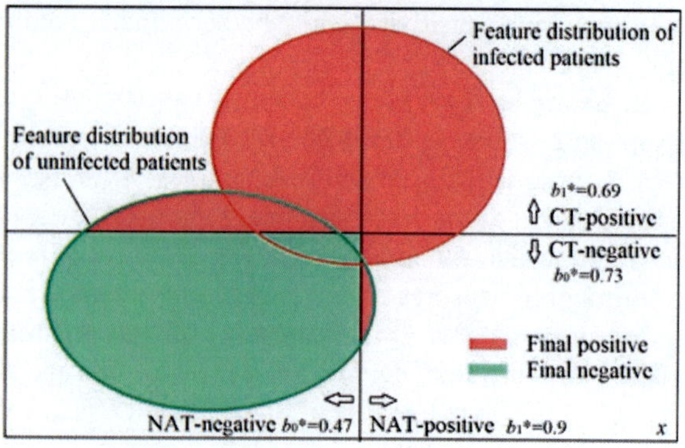

Fig. 4.10 Using both NAT and CT to diagnose the infection of COVID-19 with the help of confirmation measure b^*

Besides NAT and CT, patients' symptoms, such as fever and cough, were also used for the diagnosis. To simplify the problem, we assumed that all patients had the same symptoms so that we could diagnose only according to the results of NAT and CT.

Reference [34] introduces the sensitivity and specificity of CT that the authors achieved. According to [33, 34] and other reports on the internet, the author of this paper estimated the sensitivities and specificities, as shown in Table 4.10.

Figure 4.10 was drawn according to Table 4.10. Figure 4.10 also shows sensitivities and specificities. For example, the half of the red circle on the right side indicates that the sensitivity of NAT is 0.5.

We use c(NAT+) to denote the degree of confirmation of NAT-positive with any measure c, and used c(NAT−), c(CT+), and c(CT−) in like manner. Then we have

$$b^*(\text{NAT+}) = [P(e_1|h_1) - P(e_1|h_0)]/P(e_1|h_1) = [0.5 - (1 - 0.95)]/0.5 = 0.9;$$
$$b^*(\text{NAT}-) = [P(e_0|h_0) - P(e_0|h_1)]/P(e_0|h_0) = [0.95 - (1 - 0.5)]/0.95 = 0.47.$$

We can also obtain $b^*(\text{CT+}) = 0.69$ and $b^*(\text{CT}-) = 0.73$ in like manner (see Table 4.11).

If we only use the positive or negative of NAT as the final positive or negative, we confirm the non-infection as NAT shows negative. According to measure b^*, if we use both results of NAT and CT, when NAT shows a negative whereas CT shows positive, the final diagnosis should be positive (see blue words in Table 4.11) because $b^*(\text{CT+}) = 0.69$ is higher than $b^*(\text{NAT}-) = 0.47$. This diagnosis is the same as the improved diagnosis in Wuhan.

Assuming the prior probability of the infection $P(e_1) = 0.25$, the author calculated the various degrees of confirmation with different confirmation measures for the same sensitivities and specificities, as shown in Table 4.12.

If there is a "No" under a measure, this measure will result in a different diagnosis from the improved diagnosis. The red numbers mean that $c(\text{CT+}) < c(\text{NAT}-)$ or $c(\text{NAT+}) < c(\text{CT}-)$. Measures D, M, and F, as well as b^*, are consistent with the improved diagnosis. If we change $P(h_1)$ from 0.1 to 0.6, we will find that measure M is also not consistent with the improved diagnosis. If we believe a test-positive or test-negative when its degree of confirmation is greater than 0.2, then D is also undesirable, and only measures F and b^* satisfy our requirements.

The above sensitivities and specificities in Table 4.10 were not specially selected. When NAT-sensitivity changed between 0.3 and 0.7, or CT-sensitivity changed between 0.6 and 0.9, it was the same that only measures D, F, and b^* were consistent with the improved diagnosis.

Measure c^* is also not suitable for the diagnosis because it reflects correctness and cannot reduce the underreports of the infection. Yet, the underreports of the infection will cause greater loss than the misreports of the infection.

4.4.3 How Various Confirmation Measures Are Affected by Increments Δa and Δd

The following example is used to check if we can use popular confirmation measures to explain that a black raven can confirm "Ravens are black" more strongly than a piece of white chalk.

Table 4.13 shows the degrees of confirmation calculated with nine different measures. First, we supposed $a = d = 20$ and $b = c = 10$ to calculate the nine degrees of confirmation. Next, we only replaced a with $a + 1$ to calculate the nine degrees. Last, we only replaced d with $d + 1$ to calculate them.

The results must have exceeded many researchers' expectations. Table 4.13 indicates that all measures except c^* (see blue numbers) cannot ensure that $\Delta a = 1$ increases $f(a, b, c, d)$ more than $\Delta d = 1$. If we change b and c between 1 and 19, all measures except c^*, S, and N cannot ensure $\Delta f/\Delta a \geq \Delta f/\Delta d$. When $b > c$, measures S and N also cannot ensure $\Delta f/\Delta a \geq \Delta f/\Delta d$. The cause for measures D and M is that $\Delta d = 1$ decreases $P(h_1)$ and $P(e_1)$ more than increasing $P((h_1|e_1)$ and $P(e_1|h_1)$. The causes for other measures except c^* are similar.

4.5 Scene Understanding: How to Further Develop Our Theory

4.5.1 To Clarify the Raven Paradox

To clarify the Raven Paradox, some researchers including Hemple [3] affirm the Equivalence Condition and deny the Nicod–Fisher criterion; some researchers, such as Scheffler and Goodman [35], affirm the Nicod–Fisher criterion and deny the Equivalence Condition. There are also some researchers who do not fully affirm the Equivalence Condition or the Nicod–Fisher criterion.

First, we consider measure F to see if we can use it to eliminate the Raven Paradox. The difference between $F(e_1 \rightarrow h_1)$ and $F(h_0 \rightarrow e_0)$ is that their counterexamples are the same, yet, their positive examples are different. When d increases to $d + \Delta d$, $F(e_1 \rightarrow h_1)$ and $F(h_0 \rightarrow e_0)$ unequally increase. Therefore,

- though measure F denies the Equivalence Condition, it still affirms that Δd affects both $F(e_1 \rightarrow h_1)$ and $F(h_0 \rightarrow e_0)$;
- measure F does not accord the Nicod–Fisher criterion.

Measure b^* is like F. The conclusion is that measures F and b^* cannot eliminate our confusion about the Raven Paradox.

After inspecting many different confirmation measures from the perspective of the rough set theory, Greco et al. [15] conclude that Nicod criterion (e.g., the Nicod–Fisher criterion) is right, but it is difficult to find a suitable measure that accords with the Nicod criterion. However, many researchers still think that the Nicod criterion is incorrect; it accords with our intuition only because a confirmation measure $c(e_1 \rightarrow h_1)$ can evidently increase with a and slightly increase with d. After comparing different confirmation measures, Fitelson and Hawthorne [28] believe that the likelihood ratio may be used to explain that a black raven can confirm "Ravens are black" more strongly than a non-black non-raven thing.

Unfortunately, Table 4.13 shows that the increments of all measures except c^* caused by $\Delta d = 1$ are greater than or equal to those caused by $\Delta a = 1$. That means that these measures support the conclusion that a piece of white chalk can confirm "Ravens are black" more strongly than (or as well as) a black raven. Therefore, these measures cannot be used to clarify the Raven Paradox.

However, measure c^* is different. Since $c^*(e_1 \rightarrow h_1) = (a - c)/(a \vee c)$ and $c^*(h_0 \rightarrow e_0) = (d - c)/(d \vee c)$, the Equivalence Condition does not hold, and measure c^* accords with the Nicod–Fisher criterion very well. Hence, the Raven Paradox does not exist anymore according to measure c^*.

4.5.2 About Incremental Confirmation and Absolute Confirmation

In Table 4.13, if the initial numbers are $a = d = 200$ and $b = c = 100$, the increments of all measures caused by $\Delta a = 1$ will be much less than those in Table 4.13. For example, $D(e_1 \rightarrow h_1)$ increases from 0.1667 to 0.1669; $c^*(e_1 \rightarrow h_1)$ increase from 0.5 to 0.5025. The increments are about 1/10 of those in Table 4.13. Therefore, the increment of the degree of confirmation brought about by a new example is closely related to the number of old examples or our prior knowledge.

The absolute confirmation requires that

- the sample size n is big enough;
- each example is selected independently;
- examples are representative.

Otherwise, the degree of confirmation calculated is unreliable. We need to replace the degree of confirmation with the degree interval of confirmation, such as [0.5, 1] instead of 1.

4.5.3 Is Hypothesis Symmetry or Consequent Symmetry Desirable?

Elles and Fitelson defined HS by $c(e, h) = -c(e, -h)$. Actually, it means $c(x, y) = -c(x, -y)$ for any x and y. Similarly, ES is Antecedent Symmetry, which means $c(x, y) = -c(-x, y)$ for any x and y. Since e and h are not the antecedent and the consequent of a major premise from their point of view, they cannot say Antecedent Symmetry and Consequent Symmetry. Consider that $c(e, h)$ becomes $c(h, e)$. According the literal meaning of HS (Hypothesis Symmetry), one may misunderstand HS as shown in Table 4.14.

For example, the misunderstanding happens in [8, 19], where the authors call $c(h, e) = -c(h, -e)$ ES. However, it is in fact HS or Consequent Symmetry. In [19], the authors think that $F(H, E)$ (where the right side is evidence) should have HS: $F(H, E) = -F(-H, E)$, whereas $F(E, H)$ should have ES: $F(E, H) = -F(-E, H)$. However, this "ES" does not accord with the original meaning of ES in [14]. Both $F(H, E)$ and $F(E, H)$ possess HS instead of ES. The more serious thing because of the misunderstanding is that [19] concludes that ES and EHS (e.g., $c(H, E) = c(-H, -E)$), as well as HS, are desirable, and hence, measures S, N, and C are particularly valuable.

The author of this paper approves the conclusion of Elles and Fitelson that only HS (e.g., Consequent Symmetry) is desirable. Therefore, it is necessary to make clear that e and h in $c(e, h)$ are the antecedent and the consequent of the rule $e \rightarrow h$. To avoid the misunderstanding, we had better replace $c(e, h)$ with $c(e \rightarrow h)$ and use "Antecedent Symmetry" and "Consequent Symmetry" instead of "Evidence Symmetry" and "Hypothesis Symmetry".

Table 4.14 Misunderstood HS (Hypothesis Symmetry) and ES (Evidence Symmetry)

	HS or consequent symmetry	*ES* or antecedent symmetry
Misunderstood HS	$c(e, h) = -c(e, -h)$	$c(h, e) = -c(-h, e)$
Misunderstood ES	$c(h, e) = -c(h, -e)$	$c(e, h) = -c(-e, h)$

4.5.4 About Bayesian Confirmation and Likelihoodist Confirmation

Measure D proposed by Carnap is often referred to as the standard Bayesian confirmation measure. The above analyses, however, show that D is only suitable as a measure for selecting hypotheses instead of a measure for confirming major premises. Carnap opened the direction of Bayesian confirmation, but his explanation about D easily lets us confuse a major premise's evidence (a sample) and a consequent's evidence (a minor premise).

Greco et al. [19] call confirmation measures with conditional probability $p(h|e)$ as Bayesian confirmation measures, those with $P(e|h)$ as Likelihoodist confirmation measures, and those for $h \rightarrow e$ as converse Bayesian/Likelihoodist confirmation measures. This division is very enlightening. However, the division of confirmation measures in this paper does not depend on symbols, but on methods. The optimized proportion of the believable part in the truth function is the channel confirmation measure b^*, which is similar to the likelihood ratio, reflecting how good the channel is. The optimized proportion of the believable part in the likelihood function is the prediction confirmation measure c^*, which is similar to the correct rate, reflecting how good the probability prediction is. The b^* may be called the logical Bayesian confirmation measure because it is derived with Logical Bayesian Inference [17], although $P(e|h)$ may be used for b^*. The c^* may be regarded as the likelihoodist confirmation measure, although $P(h|e)$ may be used for c^*.

This paper also provides converse channel/prediction confirmation measures for rule $h \rightarrow e$. Confirmation measures $b^*(e \rightarrow h)$ and $c^*(e \rightarrow h)$ are related to misreporting rates, whereas converse confirmation measures $b^*(h \rightarrow e)$ and $c^*(h \rightarrow e)$ are related to underreporting rates.

4.5.5 About the Certainty Factor for Probabilistic Expert Systems

The Certainty Factor, which is denoted by CF, was proposed by Shortliffe and Buchanan for a backward chaining expert system [7]. It indicates how true an uncertain inference $h \rightarrow e$ is. The relationship between measures CF and Z is $CF(h \rightarrow e) = Z(e \rightarrow h)$ [36].

As pointed out by Heckerman and Shortliffe [36], the Certainty Factor method has been widely adopted in rule-based expert systems, it also has its theoretical and practical limitations. The main reason is that the Certainty Factor method is not compatible with statistical probability theory. They believe that the belief-network representation can overcome many of the limitations of the Certainty Factor model; however, the Certainty Factor model is simpler than the belief-network representation; it is possible to combine both to develop simpler probabilistic expert systems.

Measure $b^*(e_1 \to h_1)$ is related to the believable part of the truth function of predicate $e_1(h)$. It is similar to $CF(h_1 \to e_1)$. The differences are that $b^*(e_1 \to h_1)$ is independent of $P(h)$ whereas $CF(h_1 \to e_1)$ is related to $P(h)$; $b^*(e_1 \to h_1)$ is compatible with statistical probability theory whereas $CF(h_1 \to e_1)$ is not.

Is it possible to use measure b^* or c^* as the Certainty Factor to simplify belief-networks or probabilistic expert systems? This issue is worth exploring.

4.5.6 *How Confirmation Measures F, b*, and c* Are Compatible with Popper's Falsification Thought*

Popper affirms that a counterexample can falsify a universal hypothesis or a major premise. However, for an uncertain major premise, how do counterexamples affect its degree of confirmation? Confirmation measures F, b^*, and c^* can reflect the importance of counterexamples. In Example 1 of Table 4.9, the proportion of positive examples is small, and the proportion of counterexamples is smaller still, so that the degree of confirmation is large. This example shows that to improve the degree of confirmation, it is not necessary to increase the conditional probability $P(e_1|h_1)$ (for b^*) or $P(h_1|e_1)$ (for c^*). In Example 2 of Table 4.9, although the proportion of positive examples is large, the proportion of counterexamples is not small so that the degree of confirmation is very small. This example shows that to raise degree of confirmation, it is not sufficient to increase the posterior probability. It is necessary and sufficient to decrease the relative proportion of counterexamples.

Popper affirms that a counterexample can falsify a universal hypothesis, which can be explained by that for the falsification of a strict universal hypothesis, it is important to have no counterexample. Now for the confirmation of a universal hypothesis that is not strict or uncertain, we can explain that it is important to have fewer counterexamples. Therefore, confirmation measures F, b^*, and c^* are compatible with Popper's falsification thought.

Scheffler and Goodman [35] proposed selective confirmation based on Popper's falsification thought. They believe that black ravens support "Ravens are black" because black ravens undermine "Ravens are not black". Their reason why non-black ravens support "Ravens are not black" is that non-black ravens undermine the opposite hypothesis "Ravens are black". Their explanation is very meaningful. However, they did not provide the corresponding confirmation measure. Measure $c^*(e_1 \to h_1)$ is what they need.

4.6 Concluded Remarks and Outstanding Questions

Using the semantic information and statistical learning methods and taking the medical test as an example, this paper has derived two confirmation measures $b^*(e \to h)$ and $c^*(e \to h)$. The measure b^* is similar to the measure F proposed by

Kemeny and Oppenheim; it can reflect the channel characteristics of the medical test like the likelihood ratio, indicating how good a testing means is. Measure $c^*(e \rightarrow h)$ is similar to the correct rate but varies between -1 and 1. Both b^* and c^* can be used for probability predictions. The b^* is suitable for predicting the probability of disease when the prior probability of disease is changed. Measures b^* and c^* possess symmetry/asymmetry proposed by Elles and Fitelson [14], monotonicity proposed by Greco et al. [16], normalizing property (between -1 and 1) suggested by many researchers. The new confirmation measures support absolute confirmation instead of incremental confirmation.

The chapter was reorganized from [37]. We have shown that most popular confirmation measures cannot help us diagnose the infection of COVID-19, but measures F and b^* and the like, which are the functions of likelihood ratio, can. It has also proved that popular confirmation measures did not support the conclusion that a black raven could confirm more strongly than a non-black non-raven thing, such as a piece of chalk. It has shown that measure c^* could definitely deny the Equivalence Condition and exactly reflect Nicod–Fisher Criterion, and hence, could be used to eliminate the Raven Paradox. The new confirmation measures b^* and c^* as well as F indicates that fewer counterexamples' existence is more important than more positive examples' existence; therefore, measures F, b^*, and c^* are compatible with Popper's falsification thought.

When the sample is small, the degree of confirmation calculated by any confirmation measure is not reliable, and hence, the degree of confirmation should be replaced with the degree interval of confirmation. We need further studies combining the theory of hypothesis testing. It is also worth conducting further studies ensuring that the new confirmation measures are used as the Certainty Factors for belief-networks.

References

1. Carnap, R. *Logical Foundations of Probability*, 2nd ed.; University of Chicago Press: Chicago, IL, USA, 1962.
2. Popper, K. *Conjectures and Refutations*, 1st ed.; Routledge: London, UK; New York, NY, USA, 2002.
3. Hempel, C.G. Studies in the Logic of Confirmation. *Mind* **1945**, *54*, 1–26, 97–121.
4. Nicod, J. *Le Problème Logique De L'induction*; Alcan: Paris, France, 1924; p. 219 (Engl. Transl. The logical problem of induction. In *Foundations of Geometry and Induction*; Routledge: London, UK, 2000).
5. Mortimer, H. *The Logic of Induction*; Prentice Hall: Paramus, NJ, USA, 1988.
6. Horwich, P. *Probability and Evidence*; Cambridge University Press: Cambridge, UK, 1982.
7. Shortliffe, E.H.; Buchanan, B.G. A model of inexact reasoning in medicine. *Math. Biosci.* **1975**, *23*, 351–379.
8. Crupi, V.; Tentori, K.; Gonzalez, M. On Bayesian measures of evidential support: Theoretical and empirical issues. *Philos. Sci.* **2007**, *74*, 229–252.
9. Christensen, D. Measuring confirmation. *J. Philos.* **1999**, *96*, 437–461.
10. Nozick, R. *Philosophical Explanations*; Clarendon: Oxford, UK, 1981.
11. Good, I.J. The best explicatum for weight of evidence. *J. Stat. Comput. Simul.* **1984**, *19*, 294–299.

12. Kemeny, J.; Oppenheim, P. Degrees of factual support. *Philos. Sci.* **1952**, *19*, 307–324.
13. Fitelson, B. Studies in Bayesian Confirmation Theory. Ph.D. Thesis, University of Wisconsin: Madison, WI, USA, 2001.
14. Eells, E.; Fitelson, B. Symmetries and asymmetries in evidential support. *Philos. Stud.* **2002**, *107*, 129–142.
15. Greco, S.; Słowiński, R.; Szczęch, I. Properties of rule interestingness measures and alternative approaches to normalization of measures. *Inf. Sci.* **2012**, *216*, 1–16.
16. Greco, S.; Pawlak, Z.; Słowiński, R. Can Bayesian confirmation measures be useful for rough set decision rules?. *Eng. Appl. Artif. Intell.* **2004**, *17*, 345–361.
17. Lu, C. Semantic information G theory and Logical Bayesian Inference for machine learning. *Information* **2019**, *10*, 261.
18. Sensitivity and specificity. Wikipedia the Free Encyclopedia. https://en.wikipedia.org/wiki/Sensitivity_and_specificity (accessed on 27 February 2020).
19. Greco, S.; Słowiński, R.; Szczech, I. Measures of rule interestingness in various perspectives of confirmation. *Inf. Sci.* **2016**, *346–347*, 216–235.
20. Lu, C. A generalization of Shannon's information theory. *Int. J. Gen. Syst.* **1999**, *28*, 453–490.
21. Shannon, C.E. A mathematical theory of communication. *Bell Syst. Tech. J.* **1948**, *27*, 379–429, 623–656.
22. Tarski, A. The semantic conception of truth and the foundations of semantics. *Philos. Phenomenol. Res.* **1994**, *4*, 341–376.
23. Davidson, D. Truth and meaning. *Synthese* **1967**, *17*, 304–323.
24. Tentori, K.; Crupi, V.; Bonini, N.; Osherson, D. Comparison of confirmation measures. *Cognition* **2007**, *103*, 107–119.
25. Glass, D.H. Entailment and symmetry in confirmation measures of interestingness. *Inf. Sci.* **2014**, *279*, 552–559.
26. Susmaga, R.; Szczęch, I. Selected group-theoretic aspects of confirmation measure symmetries. *Inf. Sci.* **2016**, *346–347*, 424–441.
27. Thornbury, I.R.; Fryback, D.G.; Edwards, W. Likelihood ratios as a measure of the diagnostic usefulness of excretory urogram information. *Radiology* **1975**, *114*, 561–565.
28. Fitelson, B.; Hawthorne, J. How Bayesian confirmation theory handles the paradox of the ravens. In *The Place of Probability in Science*; Eells, E., Fetzer, J., Eds.; Springer: Dordrecht, German, 2010; pp. 247–276.
29. Huber, F. What Is the Point of Confirmation? *Philos. Sci.* **2005**, *72*, 1146–1159.
30. Carnap, R.; Bar-Hillel, Y. *An Outline of a Theory of Semantic Information*; Technical Report No. 247; Research Lab. of Electronics, MIT: Cambridge, MA, USA, 1952.
31. Crupi, V.; Tentori, K. State of the field: Measuring information and confirmation. *Stud. Hist. Philos. Sci.* **2014**, *47*, 81–90.
32. Lu, C. Semantic channel and Shannon channel mutually match and iterate for tests and estimations with maximum mutual information and maximum likelihood. In Proceedings of the 2018 IEEE International Conference on Big Data and Smart Computing, Shanghai, China, 15 January 2018; IEEE Computer Society Press Room: Washington, DC, USA, 2018; pp. 15–18.
33. Available online: http://news.cctv.com/2020/02/13/ARTIHIHFAHyTYO6NEovYRMNh200213.shtml (accessed on 13 February 2020).
34. Wang, S.; Kang, B.; Ma, J.; Zeng, X.; Xiao, M.; Guo, J.; Cai, M.; Yang, J.; Li, Y.; Meng, X.; et al. A deep learning algorithm using CT images to screen for Corona Virus Disease (COVID-19). *medRxiv* **2020**, https://doi.org/10.1101/2020.02.14.20023028.
35. Scheffler, I.; Goodman, N.J. Selective confirmation and the ravens: A reply to Foster. *J. Philos.* **1972**, *69*, 78–83.
36. Heckerman, D.E.; Shortliffe, E.H. From certainty factors to belief networks. *Artif. Intell. Med.* **1992**, *4*, 35–52.
37. Lu C. Channels' Confirmation and Predictions' Confirmation: From the Medical Test to the Raven Paradox[J]. Entropy, 2020, https://doi.org/10.3390/e22040384.

Chapter 5
Reconciled Interpretation of Vision, Touch and Minds

Abstract This chapter aims to present a theoretical framework on the evolution stages of the machine brain and cognitive computation and systems for machine computation, learning and understanding. We divide AI subject into 2 branches—pure AI and applied AI (defined as an integration of AI with another subject: geoAI as an example). To stretch the continuation of Chap. 1, we first analyze how to predict dangers in unmanned driving with geoAI and introduce the robot path planning (RPP) problem. Subsequently, an ant colony optimization (ACO) algorithm for solving the RPP problem are interpreted to understand cognitive computation and systems for machine computation, learning and understanding. A practical example of RPP problem—the traveling salesman problem (TSP) is further introduced. Integrating ACO with the iteration-best pheromone update rule, the ACO algorithm is improved and an adaptive mechanism are presented to treat instability. Experiments show that the new ACO algorithm has a good performance and robustness. Stability of the cognitive system and its robustness in cognitive computation for solving TSP are further validated. The vision-brain hypothesis, which has been proposed in the book "Brain-inspired intelligence and Visual Perception", is developed and hence extended as the vision-minds brain hypothesis. At the end of this chapter, as a first theoretical utilization of the vision-minds brain hypothesis, we explain how artificial improvements of the algorithms in applied AI can contribute to evolution of the machine brain.

5.1 Background

Until now, brain-inspired intelligence is still a concept in debate and even, scientific interpretation of machine intelligence is still not uniform. This can be largely attributed to misunderstanding of the evolution law of the machine brain. A theoretical framework on development stages of AI and the current level of machine intelligence are still required for cognitive computation and systems in the machine brain. Over the past decades, the development of AI have experienced three stages—machine computation, machine learning and machine understanding.

W. Wang et al., *Interdisciplinary Evolution of the Machine Brain*,
Research on Intelligent Manufacturing,
https://doi.org/10.1007/978-981-33-4244-6_5

The three development stages of AI also represent the current level of machine intelligence—cognitive computation.

Scientists have taught machine how to collect and treat data and discover knowledge from the data. Evolution of machine brain will experience another two stages—machine meta-learning (learning to learn) and self-directed development (improving the capability of machine brain utilizing the learned knowledge). AI will be a driving force for the future world.

As a driving force for the future world, AI is presenting a series of tools to tackling open challenges in devious subjects. This motivates us to divide AI subject as two branches—pure AI and applied AI, where applied AI is defined as an integration of AI with other subjects. In Chap. 1, we have learned that geoAI is an integration of geospatial science with intelligent algorithms and so geoAI is applied AI. Benefitting from cognitive computation and systems in geospatial science, geoAI not only represents AI methods involved in the basic geographic information technology, but also presents the methods and solutions for intelligent classification, prediction and analysis of geospatial data in other research field associated with applications of AI.

Cognitive systems in geoAI can be developed from experiences of human. For instance, ambulances drivers can give real-time feedbacks to the changes in surrounding environments. Accuracy in driving decisions majorly depend on their experiences on traffic conditions, peaking time and so on. Skilling drivers know well about how to collect, sort and analyze necessary geospatial data for improving the accuracy of their decisions. Internet of things (IoT) can help us to learn experiences from thousands of drivers and establish precision models for cognitive computation. This is only an example. Other experiences of human in utilizing geospatial data also certainly make sense for developing cognitive systems in geoAI and in turn, geoAI automates machine learning on geospatial data and breaks limitations of pure AI in facing real challenges.

Objectives of this chapter are: (1) to present a theoretical framework on the evolution stages of the machine brain and cognitive computation and systems for machine computation, learning and understanding, (2) to interpret cognitive computation and systems for machine computation, machine learning and machine understanding, employing the robot path planning (RPP) problem as a practical example, and (3) to explain how algorithms improvements in pure AI and applied AI can contribute to interdisciplinary evolution of the machine brain, utilizing the ant colony optimization (ACO) algorithm as an example.

5.2 Preliminaries of Cognitive Computation

The development of AI have experienced three stages although, until now, scientific interpretation of machine intelligence is still not uniform. In order to interpret evolution law of the machine brain, we will first interpret the current and future development stages of machine intelligence. Three development stages of AI also represent the current level of machine intelligence—cognitive computation. Machines for environmental sensing help us to acquire the data and machine leaning has become a major tool of data mining and knowledge discovery. Benefiting from the rapid development of AI, such as pattern analysis, scene understanding and etc., machines are already getting to understand the real world.

5.2.1 Evolution Stages of the Machine Brain

Despite of the debate on whether machine will have minds (or whether artificial human will be produced) in the unknowable future, human society is already moving from information (electrical) society to intelligent society. In the future, development of machine intelligence will experience another two stages—machine meta-learning (learning to learn) and self-directed development (improving the capability of machine brain utilizing the learned knowledge). These are also evolution stages of the machine brain.

Now we will further interpret the basic evolution law—interdisciplinary evolution of the machine brain. In the current and next development stage of machine intelligence, the social needs to novel sciences and technologies are motivating a potential evolution of the machine brain. In order to present a preliminary description of the basic evolution law—interdisciplinary evolution of the machine brain, we need to reconsider a significant question—how to differentiate machine intelligence and AI? Obviously, machine intelligence is not the whole story of AI. In the famous Turing test, a machine is defined to be intelligent if it could communicate with the human through telex devices, without being able to identify its machine identity. Based on this definition, intelligent machines should have brain-inspired intelligence and even, should have minds. The real machine intelligence represents the highest stage in the development and applications of AI. The natural way to approach this highest stage requires an interdisciplinary evolution of the machine brain—to develop learning systems through interdisciplinary researches.

For example, GeoAI as an integration of geographic information system (a basic cognitive system) with machine computation, learning, understanding would potentially motivate evolution of spatial intelligence in the "machine brain". Solutions of some significant problems—how many people can the earth support, how will the ecosystems respond to global warming and etc., are related to scientists who are engaged in geographic information research. Especially in 5G era, integrated network communication satellites, navigation satellites and remote sensing satellites will further present convenience for the earth observation of space and space. There will be plenty of sensor networks and real-time dynamic geographic information to process. Reanalyzed data will further provide scientific convenience for agriculture, education, environment and etc. It is no doubt that geoAI will be a significant part of intelligence in the machine brain to analyze and process such big spacial data.

5.2.2 Cognitive Systems in GeoAI for Unmanned Driving

Novel cognitive computation and systems are essentially significant for the next evolution stages of the machine brain. Benefiting from the novel systems, machine will learn to learn and will have the ability improve the capability of machine brain utilizing the learned knowledge.

To stretch the continuation of Chap. 1, we further analyze how to predict dangers in unmanned driving with geoAI and introduce the RPP problem. This is also a preliminary attempt to construct the novel cognitive systems. In order to help the future machines to expand insights into the real world, geoAI requires more efficient AI methods for data collection, selection, transmission, storage and processing. This further motivates the development of pure AI. Moreover, these AI methods are not the whole story of cognitive systems in geoAI and in turn, it is also essentially important to develop new theory and methodology for applied AI. These new theory and methodology can be intelligent, but they will be relatively independent of the current theory and methodology for pure AI. We will further explain the above opinions through a perspective problem—how to predict dangers in unmanned vehicles?

A direct answer is to equip these unmanned vehicles with particular devices for machine vision and to design suitable algorithms for fully utilization of the collected data for geoAI. These data are of multi-source, including geospatial data, information acquired from remote sensing and real-time images/videos collected by the vision system of unmanned vehicles. Integration of geoAI and edge computing will greatly changed the utilization of the multi-source data. In order to generate a real-time feedback, the vision system of unmanned vehicles must be developed enough to collect geospatial images and videos of high-resolution. Developed vision systems also make advantages for a better function of the whole unmanned driving system.

GeoAI is an emerging branch of applied AI, but its influences will expand rapidly. It might be a key to the fourth industrial revolution! Exactly, geoAI has made advantages for us to rapidly develop the Internet of things and track behaviors of consumers, governments and enterprises. Integrating with edge computing, machine learning and the theory of blockchain, geoAI already has a wide impact on our lives (insurance, agriculture, national defense and etc.). It is quite necessary to develop the theoretical framework of applied AI and especially geoAI. Methods in AI have promoted the technological revolution in industry and will certainly change the world future, but geoAI is still in its infancy. Facing the development opportunities, the theory and methodology of geoAI can be established and developed.

Theory and methodology of geoAI will improve the machines ability of analyzing problems in the real world. Taking unmanned driving as a practical example, geoAI in the machine brain will help unmanned vehicles to work well. Cognitive systems in geoAI are of high-performance in making better decisions on self-driving. Such decisions include the optimal path selection, sub-path planning, dangers prediction and beyond. A hypothetical framework to cognitive systems in geoAI for unmanned driving is shown in Fig. 5.1.

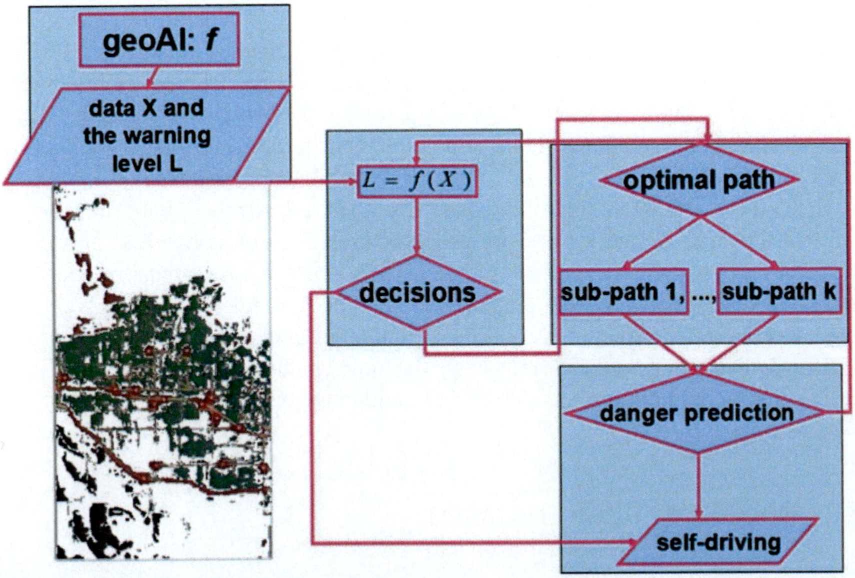

Fig. 5.1 A hypothetical framework of cognitive systems in geoAI for unmanned driving

The current achievements in computer vision, remote sensing, atmospheric science, data science, electronic and electrical engineering and especially the wide applications of geographical information system have presented a deep foundation for developing geoAI theory. Integration of geoAI with unmanned driving can be further extended to next generation of automation with geoAI. This is a very excited dream and worthy of more attention. Using geospatial data, scientists will break some limitation for technological advances in next generation of automation. GeoAI can also make advantages for the intelligent machines to build, access and run complex cognitive systems for processing and analyzing geospatial data. Such cognitive systems can help machines to extract valuable information from multi-source data by cognitive computation. Next generation of automation with geoAI will make significant progresses in deep analysis of a variety of data sources. Such deep analysis of multi-source data implies a huge business scale and a level of commercialized thinking.

5.2.3 The Robot Path Planning (RPP) Problem

Ambulanceye proposed in Chap. 1 is right a novel cognitive system in geoAI for unmanned driving of the future ambulances. Ambulanceye aims to safely detect eigenobjects around intelligent ambulances and timely predict the potential dangers.

Correct prediction means that the intelligent ambulances will perform the right operation at the right time, taking into account all the actions of eigenobjects around. Wrong predictions mean that potential dangers will happen. As artificial humans, intelligent ambulances need use thousands of environmental inputs to intuitively perform decisions, including the sub-paths planning, an optimal path selection, dangers prediction and beyond.

Since these intelligent ambulances are hypothesized as robots with minds (i.e., artificial humans). Hence, the decision process can also be summarized as a process to solve the robot path planning (RPP) problem. Current machine computation, learning and understanding largely rely on general intelligence—real-world driving experience. But the RPP problem is very difficult and more intelligence can be developed in the machine brain. At the current stage, the brain can be interpreted as a cognitive system and such intelligence is carried out through cognitive computation.

5.3 The Minds Brain Hypothesis

This section aims to propose the vision-minds brain hypothesis [1, 2]. We will utilize improved ant colony optimization (ACO) algorithms to solve the RPP problem and explain the cognition processes based on the vision-minds brain. Exactly, these cognition processes contained in ACO algorithms can also be used to interpret evolution of the machine brain from the vision brain to the vision-minds brain. RPP problem in Sect. 5.2.3 is naturally employed for a theoretical and preliminary interpretation.

5.3.1 From Vision to Vision-Minds

In Ref. [2], the vision–brain hypothesis was illustrated in three steps.

First, it was hypothesized that the machines' attention is largely driven by visual information and the attention could be regulated with brain-inspired objects detection and tracking. This can be easily understood since machine vision has become a mature research field. It is not difficult to construct a machine vision system. Intelligent algorithms in the system have a significant implication to cognition accuracy and response speed of the machine brain.

Second, it was hypothesized that if a region of interest (ROI) had been located by the vision–brain of an intelligent machine, then the machine can smoothly understand and partition the visual scene. This further help machine vision systems to improve both cognition accuracy and response speed.

Third, it was hypothesized that in the future, cognition rates of the vision brain could approach to 100%. This finally establishes the robustness and efficiency of the vision brain. A single closed loop of the current learning systems is still very simple, but when the multiple closed loops are connected with other hybrid adaptive algorithms, the integrated intelligence in the vision brain of the future robots will be powerful.

After the human obtain information through visual perception, they always have minds (a force of habit) to utilize the visual data. At the current stage, intelligence of a single machine brain is still very simple and largely depends on AI. Each small closed loop of the machine vision systems can be taken as a small ant. But in the unknowable future, the huge intelligent machines colony are bound to have great wisdom. Theoretically, it is feasible to develop the vision brain hypothesis and extend it as the vision-minds brain hypothesis.

5.3.2 *Understanding the Minds Brain Hypothesis*

ACO is originated from biological studies, where it was demonstrated that ants are deposit pheromone on the ground while walking. They can collect and exploit these pheromone to find an optimal path between their nest and a food source. Amount of pheromone are proportional to the quality of the final choice [1]. Because ants have little vision, these cognition processes majorly depends on smell, touch and minds. Therefore we will employ these cognition processes in ACO to interpret the minds brain hypothesis.

Extending the explanation from intelligent ambulances to general robots, the task of the cognitive system can be divided into a series of subtask. Each subtask corresponds to cognition of a subpath. The full path is generated by integrating all the subpaths.

This is not only our main idea to improve the ACO algorithm, but also the main idea to solve the RPP problem. Exactly, ACO has become a typical algorithm for interpreting machine computation, learning and understanding. It is a population-based metaheuristic system for cognitive computation. In ACO, each individual of the population can be taken as an artificial ant. These artificial ants are born incrementally and stochastically. They move step-by-step on a graph-based representation of the RPP problem. Each step of their moves define a component of the solution. These solution components will generate the final solution of the RPP problem.

Evolution of the machine brain based on these rules is as follows. Taking into account bias of ants in their decisions, a probabilistic model is introduced and updated on-line by current ants to increase the probability of the future ants to build better solutions. Similar to real ants, artificial ants can deposit pheromone. Each subpath is represented by a pheromone trail. Alternatively, decisions are made on the graph of the RPP problem by counting the crossed ants of each subpath. ACO is always combined with the iteration-best update rule—a modification rule for the pheromone iteration. This is a subpath-based rule to enhance the pheromone depositing strategy [2–5].

5.3.3 *The Traveling Salesman Problem (TSP)*

In formulation of the cognition processes and solution of the well-known RPP problem, we will focus on a practical example of the RPP problem—the traveling salesman problem (TSP): given a list of cities and the distances between each pair of cities, what is the shortest possible route that visits each city and returns to the origin city?

TSP is a classical NP-Complete problem [4]. As a meta-model of many problems in reality, TSP has received a wide attention and a series of ACO algorithms have been proposed to solve TSP [1–7]. TSP is one of the most studied discrete optimization problem with a large number of applications. It is easy to formulate, but it is difficult to solve.

It is easy to see that cognition in ACO majorly depends on pheromone updating models. There are two forms of cognition—one associates the pheromones with arcs and the other associates pheromones with nodes [8–12]. As seen in Sect. 5.3.1, the task of the cognitive system has been divided into a series of subtask. Each subtask corresponds to the ACO cognition of a subpath. The full path is generated by integrating all the subpaths. We symbol the ACO algorithm with this subpath-based pheromone modification rule as SPB-ACO algorithm and improve the ACO cognition process. The solution of TSP is carried out by combining the SPB rule and the iteration-best update rule, an adaptive SPB-ACO is used to solving standard TSP.

The SPB-ACO cognition processes contains two components:

1. iterative updating process—count the crossed ants of a subpath in each iteration and select r subpaths to update.
2. adaptive process—regulate the updating strength with the iteration.

Combining with the iteration-best update rule, the major step in SPB-ACO cognition process is repeatedly selecting cities until performing a complete solution and calculate the ants' fitness to find an optimal route.

A solution of TSP based on the above rules has been presented in [1].

5.4 Cognitive Computation and Machine Minds

In this section, cognitive computation and systems in minds of ant colonies are theoretically summarized through a brief review of some representative studies and applications on ACO algorithms.

For a further understanding of interdisciplinary evolution of the machine brain and a preliminary illustration of the relationship between cognitive computation and machine minds, we introduce another example of applied AI—medAI (an integration of AI with medical science).

Exactly, geoADAS for the future of medical rescues proposed in Chap. 1 can also be recognized as a practical application of medAI. In this section, design of percutaneous robots (DPR) will be employed as another practical application of medAI. DPR is also utilized to further analyze the vision-minds brain hypothesis and its cognitive models.

5.4.1 *Supplementary Explanation of Machine Minds*

It is easy to see that ants colonies as social organisms have their collective intelligence. Such collective intelligence can also be interpreted as collective minds of these ants colonies. There are plenty of studies on the collective intelligence and its enlightening significance to the theoretical research of ACO algorithms [5, 13–22]. Particularly, distributed optimization processes by ant colonies, computational models of rank-based version of the ant system and the hyper-cube framework for ant colony optimization are studied [13, 15, 20]. Collective intelligence of ants colonies has been cultivated in the machine brain for a better solution of TSP [5, 14, 16–19]. Pheromone mark ACO with a hybrid node-based pheromone update strategy was constructed [22] and the strategy (or rules) to function the collective intelligence of the ants colonies are can be recognized as a significant part of their collective minds. Therefore, it is possible and feasible to cultivate these collective minds in the machine brain [5, 21–25]. Integration of ACO with reinforcement learning systems and other brain-inspired models can be taken as preliminary attempts to approach such cultivation of machine minds [3–6, 21]. It is true that machine minds are still a dream for the unknowable future, but it is also true that effectiveness and efficiency of ACO algorithms have been greatly improved for a better solution of the RPP problems [7–12, 26, 27].

Since it is easy to construct a machine vision system for robots, we can also further develop ACO algorithms based on the vision-minds hypothesis. We not only will teach machines to learn collective intelligence of ants colonies, but also will teach machines to integrate the collective minds with a visual perception [8, 28–32]. Improved ACO algorithms will be applicable to solve other problems the real world—multi-target tracking, multi-agent dynamic manufacturing scheduling and etc. [8, 9]. In previous studies, vision-based navigation system by variable-template matching and self-organized shortcuts in ACO has been developed for autonomous mobile robots [28, 30–32].

5.4.2 *From Machine Learning to Machine Understanding*

It has been widely recognized that the current evolution of the machine brain largely depends on the development of AI—especially depends on the evolution of machine learning methodology. With the rapid development of machine learning methodology, deep learning systems, reinforcement learning systems and other brain-inspired learning systems have been successfully applied to many research fields.

Besides geoAI and medAI, integration of AI with another subject is also feasible. The era of applied AI is on the way!

Effective training of these machine learning systems not only requires the professional knowledge of the corresponding subjects, but also requires some significant experiences in artificial understanding and solutions of concrete issues. Development of the corresponding subjects and updates of artificial experiences are potentially driving interdisciplinary evolution of the machine brain. Machine intelligence is developing from machine learning to machine understanding—a new stage to approach machine minds. In order to further understand interdisciplinary evolution of the machine brain and also present a preliminary illustration of the relationship between cognitive computation and machine minds, design of percutaneous robots (DPR) will be employed as another practical application of medAI in Sect. 5.4.3 to further analyze the vision-minds brain hypothesis and its cognitive models.

5.4.3 The Vision-Minds Brain Hypothesis and Associated Cognitive Models

DPR is based on the infrared vision of vascular localization technology [33–35]. Vascular localization has been successful in clinic, but automatic control of the syringe needle in DPR is still a great challenge [36–40]. We also need to take into account differences among individual patients [41, 42]. Automatic control of the syringe needle depends on intelligent needle navigation, which is bound to become a key direction of DPR [43–47].

The vision-minds brain hypothesis will present an ultimate solution to DPR, where the robot will work as a professional nurse. They consciously decide whether it works or not, and if it doesn't work, the choice will be repeated [48]. In order to training cognitive models associated with the vision-minds brain hypothesis for DPR, a large number of correct venipuncture operations are required [49]. This is obviously a complex cognitive computing problem.

Based on the vision-minds brain hypothesis, DPR can be carried out with image processing and automatic control technology. The hypothesis will make sure that the clinic operation is reliable and stable. For vascular localization, the machine vision system is mainly composed of industrial control computer, infrared light source, infrared industrial camera and red external filter. The vessel localization technology can be realized with the help of infrared vision. Denoising results and the vascular localization effect are shown in Fig. 5.2.

The key problem is how to push the needle to the right puncture point. This can be taken as a 3D path planning problem—a difficult RPP problem. The robot not only need to achieve complete intelligent needles of navigation and control with the help of a three-dimensional path planning, but also need to carry out an injection operation according to the vein diameter and depth.

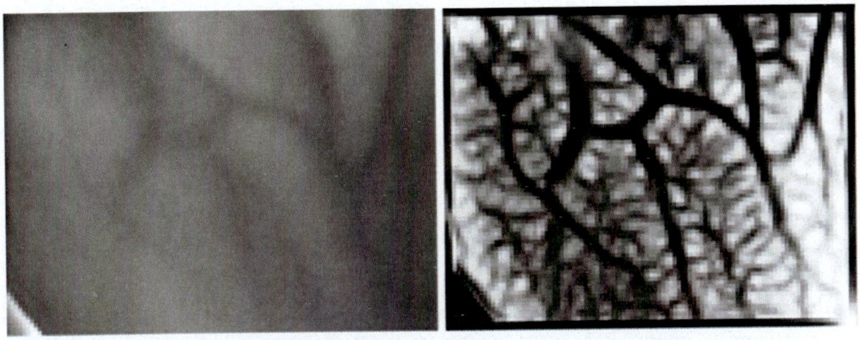

Fig. 5.2 Denoising image and the result of vessel location

5.5 Reconciled Interpretation of Vision, Touch and Minds

5.5.1 Improvements of the Cognitive System

As a first theoretical utilization of the vision-minds brain hypothesis, we explain how artificial improvements of algorithms in applied AI contribute to evolution of the machine brain. Improvements of the medAI cognitive system for DPR is naturally employed for illustration.

The vision-minds brain hypothesis is theoretically utilized in a preselection of intravenous injection points. As seen in Fig. 5.3, the main idea to select the injection point G is as follows. The vision system decides the first assessment of the major blood vessels. Then the robot has their own minds to make a final decision by comparing the local average width of these blood vessels.

The cognitive system can be further improved by introducing processes of error determination and correction for efficient decision. This helps the robot to identify impending operational errors before they are punctured and correct them in advance.

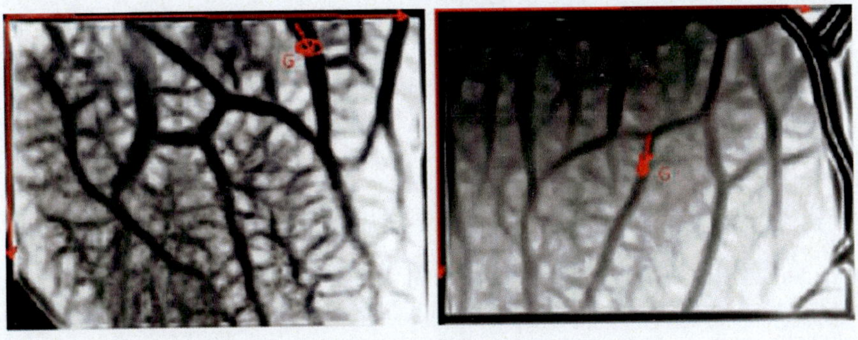

Fig. 5.3 Right venipuncture points based on the vision-minds hypothesis

Consequently, artificial improvements of algorithms in applied AI contribute to evolution of the machine brain.

Brain-inspired computing and brain-inspired intelligence emphasize the combination of brain science and neuroscience. After the emergence of deep learning, this kind of combination becomes more important in the field of AI. The vision-minds hypothesis will enables medAI researchers to review and summarize past research achievements in DPR and other aspects from the perspective of deep learning, and realize the cognitive system in the future. To realize the dream, we still need to established suitable models for cognitive computation and construct automatic systems for the intelligent control and decision. The complexity and variability of the clinical intravenous injection processes need to be taken into account. This is a great challenge.

5.5.2 *Further Interpretation of the Vision-Minds Hypothesis*

In this section, we provide possible answers towards the questions about where human intelligence and subjective consciousness come from, which can be concluded as the Skin Brain Hypothesis [50]. It reveals the origin of human intelligence and the fundamental differences between human and machines, and also provides a basis for our study on breaking through the bottleneck of machine intelligence by proposing the vision-minds hypothesis.

Nowadays, artificial intelligence has surpassed humans in many professional fields. For example, AlphaZero can beat the champion of human chess players, AlphaFold can beat the scientists and experts in predicting protein folding, and AlphaStar can beat professional players in competitive computer games, and so on. If Moore's Law continues to work in the future, by 2045, the performance of machines will be 10^{26} times that of the original, which is more than the combination of all humans. Today, chip technology is on the nanosecond scale, and biotechnology is on the nanometer scale. In contrast, humans react only in milliseconds without tools, and the scale of perceived change is in millimeters, which are six orders of magnitude different from machines.

The general public has not enough understanding of the current situation and development trend of AI and therefore form into two extremes.

The first point of view, that AI is just a mathematical model, is undoubtedly a narrow perception. In many ways, machines have acquired capabilities beyond human control. For example, deep learning based on model structure has shown excellent ability in multiple tasks, and its learning process will make the model develop towards the direction that cannot be fully explained by human beings, which is undoubtedly inexplicable by the static model theory.

The second view is that artificial intelligence will become some organism like a human being, or even making it impossible for us to tell the difference between AI and humans. This view, especially from the philosophical point of view, falls into the misunderstanding of physical reduction and the misunderstanding of human intelligence. Human intelligence is unique, and if artificial intelligence continues to develop along the current path, it will be difficult to simulate human-like intelligence successfully.

Are machines able to eventually simulate human intelligence, surpass human intelligence, and replace human intelligence? There are two different mainstream views. One is that human have transcendence while Turing machines do not. The other is that a human is a machine made up of atoms and which can be physically restored [51–53]. We assume that the key lies in whether machines can develop human-like consciousness or not. We believe that human beings have subjectivity and are able to deal with fuzzy problems at will, and quickly pass over the possible dark infinity from the perspective of "sensibility" rather than "rationality". This is a very difficult ability to have for the present artificial intelligence.

The boundary between intelligence and non-intelligence is difficult to distinguish, whether in terms of thinking ability or learning ability, it is hard to define the essential difference between humans and machines. And even in the nature, intelligence also has a variety of forms. If we admit that animals, such as dogs, are intelligent, then we cannot deny that present AI has a certain degree of intelligence already, and this intelligence is still on the way to simulate human. The "Chinese room" problem is often used to prove that machines have no intelligence, but in the process of constructing dictionaries in which humans condensate and transmit consciousness to machines. The intelligence of machines is not evolved by itself, but given by human beings, and can be described as goal-oriented rather than cognitively oriented. From the perspective of application, this will lead to the problem that the machine cannot solve the boundary ambiguity, while from the perspective of intelligence, the defect of the machine is that it cannot have subjectivity.

For human beings, the subjectivity derives from the strong sense of telling Self from the Outside World. In 0–5 years old stage, brain develops very fast, based on which we propose the Skin Brain Hypothesis. It implies that gene mutations leading to a reduction of hair, so that human skin has become sensitive, which provides a significant physical boundary between the human body and the outside world. In the rapid development stage of the brain, neurons are quickly connected by strong stimulations from the skin, such as cold, pain, etc., which makes the baby have a distinction between the awareness of Self and Outside World. This distinction of Self and Outside World evolves into the final subdivision of relevant concepts, shaping into a strong sense of self-awareness. It is the largest regional difference in humans and other animals, but also a core factor. We believe that the sense of touch facilitates the generation of feeling being an individual, which is much like an attractor in Physics. Once produced, it is not easy to disappear, thus the sense of touch is the most important factor to evolve advanced intelligence. Zoologists raise the obstetrical dilemma [54], whose explanation is that because a woman's pelvis is not big enough, so the infant must be delivered before its brain getting mature.

In our opinion, this is just conducive to the growth of baby's self-awareness, that is when the neural connections of the brain are not fully formed, and it is able to accept external stimuli to generate a strong sense of self-awareness. Merely having a sensitive skin is not enough for advanced intelligence, the life also needs to survive in harsh environment, so the generation of advanced intelligence needs very strong conditions, which also helps us to understand the Fermi Paradox, because advanced intelligence is actually very rare from the perspective of evolution.

The proto-consciousness refers to the final subdivision of individual Self and Outside World, which is the initial condition to generate all intelligence, and I think, therefore I am proposed by Rene Descartes also shows that Self exists without doubt. There are many quale, but the quale that can be perceived are very limited indeed. Linguists once sums up how the various colors appear in our expressions [55], and they find that if there are only two colors, they must be related to black and white; if there are three, they must be black, white, and red; if there are five, it must be black, white, red, yellow and green. That means the strong concept of color is differentiated from the simple black and white, which also gives us a clue to know how people understand the world. Concepts may initially be very vague, which will be more accurate after constant iterations (between individuals and generations).

5.5.3 Extension of Boundaries and the Skin Brain Hypothesis

The reduction of hairs and the sensitivity of the skin caused by genetic mutations provides a clear physical boundary between the human body and the outside world, as well as a physical basis for the proto-consciousness. With the rapid development of the brain, connecting and strengthening the neurons connections, the proto-consciousness forms a strong sense of Self, in order to further explore the boundary between Self and Outside World, and thus produces advanced intelligence. The whole evolutionary process is defined as the Skin Brain Hypothesis, as shown in Fig. 5.4.

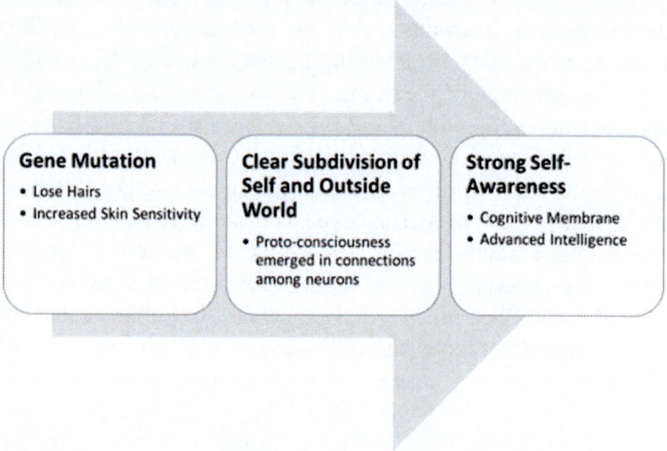

Fig. 5.4 The process of skin brain hypothesis

The Skin Brain Hypothesis makes it clear that touch provides a physical basis for distinguishing between Self and Outside World and thus has a significant place in the evolution of human intelligence. The proto-consciousness is the starting point of individual cognition, which can transmit between individuals and generations. Cognitive membrane contains the concept system, value system and belief system of human beings, as the individual cognition is deepening, and the cognitive membrane is expanding, and it provides protection for the growth of Self. Like the cell membrane to protect the nucleus, the cognitive membrane plays a role in protecting self-awareness, it on the one hand filters the outside information, and select useful parts into the main cognitive system, on the other hand in the face of external pressure, it subjectively narrows the gap between the Self and Outside World, so that the individual can maintain a positive attitude.

For individuals, self-awareness and cognitive ability are synchronized, and human's self-awareness begins to evolve from birth. Self-awareness is rooted in the individual's sensory system. After the baby is born, the primitive sense of touch (which directly produces strong stimuli such as warmth and pain), hearing and sight enable him to distinguish himself from the outside world, making him develop the simplest sense of self. We propose the skin brain hypothesis, which points out that the hairlessness and skin sensitivity caused by genetic mutation in the evolution of human beings provide a clear physical boundary between the human body and the outside world, and also provide a physical basis for the division of Self and Outside World.

In the evolutionary view for the human beings, touch is more important. Buddhists say that "eyes, ears, nose, tongue, body and mind", and the "eyes" come first, standing for vision. Vision and the endocrine system are less closely linked to the emergence of consciousness than touch, so babies are less likely to distinguish themselves from the outside world only by vision ability. Many organisms have highly developed visual systems but do not have the senior intelligence as humans, perhaps because visual, olfactory and other stimuli are not easy to distinguish the Self of the organism from the Outside World. We believe that touch is conducive to the generation of individual sensations and an important factor for the generation of advanced intelligence, which is like the attractors in physics, that are not easy to disappear once emerged. When the proto-consciousness has been generated, then for the development of individual intelligence, or for the machine that already has fragments of consciousness, the influence of touch is no longer prominent, and the vision gradually occupies a dominant position.

Turing machine is driven by an algorithm, but there are still problems in the current theoretical algorithm. Human intelligence evolution and human subjective bias is intertwined, if our intuitive system perfectly reflects all the outside world at the beginning, we really do not need the intelligence. Like the Almighty God, he knows everything and needs neither intelligence nor language, and he only needs to query like a machine. The Turing machine itself cannot produce self-awareness or value systems. If present Turing machines and human consciousness are combined together, then the new machines can be considered to have awareness. As there is a reward system in the Reinforcement Learning of AlphaGo, which actually provides a purpose and can be seen as a simplified value system, AlphaGo have certain awareness to some extent. In the future, to make the machines evolve human-like intelligence, one possible path is to repeat or simulate human evolution, so that the machine can perceive the world. That idea is just like the Internet of Things technology, providing machines with feeling units, similar to our tactile function of skin.

The significance of artificial intelligence is not to form a general system or AGI to solve all problems, but to see the importance of subjectivity in intelligence. If we want to achieve qualitative breakthrough in AI development, then we must give machines subjectivity instead of making them passively adapting to the outside world. Cognitive agents and object constraints are usually inconsistent at first, and then the agents can reach consensus or compromise with physical world through cognitive attractors or fragments of consciousness. At the beginning, cognitive agents were much restricted by the physical world. With the evolution proceeds, subjectivity gradually become strong and dominant. If there is not any ability to detect, process and use cognitive attractors, then there is not intelligence, which is applicable for both humans and machines.

This section was developed from [50, 56–58]. With the basis of studying the fundamental differences between human and machine intelligence, it is clear that the key to open up the rapid development of strong AI in the future depends on how to make machines gain strong self-awareness. Deep learning offers the possibility of implementing continuous interactions between machines and the outside world, but the problem is that it is difficult for machines to recognize the boundaries, which are very obscure and significant as the starting point of human consciousness. Now that the skin brain hypothesis can explain the origin of human intelligence, does that means machines should have started with the sense of touch as well? Our answer would be to negative. As for machines, the critical point will be vision.

Unlike carbon-based human beings, which survived out after thousands of years of natural evolutions, silica-based machines are created by humans and have acquired fragments of consciousness from us already. Touch is more important at the beginning of consciousness for making the individual recognize oneself out of the environment, while in the later stage of developing intelligence, vision becomes more and more important. To match the computing power and its cognition level, it is feasible to train a machine with vision data, which is more abundant, easier to process and better localized. The vision-minds hypothesis can therefore plays a vital role in improving machine intelligence.

5.6 Simulation of the Transponder Transmission System

Hypothesize that the future machines already had a vision-minds brain, how to construct the transponder transmission system for efficient decisions? This section will theoretically explain how to construct a computer simulation system on transponder transmission, utilizing the intelligent system in the train for transponder transmission as a practical example.

Simulation technology has been widely used in a series of scientific research—engineering design, military exercises and online algorithms updates for more efficient machine sensing. Computer simulation technology is formed by the combination of modern computer technology and simulation. The marriage of computer and simulation technology can be said to be a boost. It opens up a broader application space for simulation technology and plays an increasingly important role in studies on machine intelligence. Transponder transmission system is the key component in the train operation control system. Besides the data from the vision-minds brain, it also transmits the basic parameters—line speed information, temporary speed limit information, special positioning information and other train control safety information to the running train through the transponder up-link signal [59–61]. So it is very important for us to master the key technology of the transponder transmission system to carry out the transponder upstream signal experiment [62–65]. Computer simulation system will present an important means and method for the vision-minds brain to analyze the system operation behavior, monitor the system dynamic process and improve the motion decision. Benefiting from the development of system science, control theory, computing technology, computer science and technology, especially from the rapid development of information processing technology in recent years, simulation technology has developed rapidly [66–70]. Integration of the vision-minds brain with simulation technology will extend a broader application space for simulation technology and plays an increasingly important role [69–72].

5.6.1 Development History of Transponders Transmission System

As early as many years ago, many railways developed point-based parking devices. When a train passes through a point equipped with such devices, if there is a stop signal, the devices immediately notify the train driver to stop the train. This kind of point-based parking device was originally mechanical, and later developed into an electronic device. Induction coils between the locomotive and the ground collect some frequency signals and send parking information. Resonance occurs between the coils of the vehicle and the roadside, which produces accurate information to the locomotive device. This kind of device is actually one of the earliest forms of query transponder, and also the initial form of application in railway. After the transponder was put forward, Germany, the Netherlands, the United States, Japan, France and so on have adopted this query transponder as the information transmission mode of the ground locomotive. Early transponders did not have uniform technical specifications, which made the type, size and technical characteristics of ground transponders used in trunk railways and urban rail transit in various European countries very different. For example, Siemens ground transponders were widely used in Germany to match the point train speed control system ZUB, while Nordic railway trunk lines and urban rail transit intersected. Tongzhong uses a large number of flat-panel ground transponders produced by a company. Until the end of 1996, the European Railway Union published the detailed technical specifications of ETCS, in which the European ground transponder was explained. The unified size standard, interface standard, type standard and reference standard of technical parameters have formed.

Because of the need of the development of high-speed railway and passenger dedicated line in China, the Ministry of Railway independently formulated the technical standard of Chinese train control system in 2001 on the basis of European train control system D, which determined the overall technical scheme and overall plan for the leapfrog development of railway signals in China. With several speedup of railway transportation in recent years, the safety information provided by previous monitoring equipment has gradually become insufficient. At this time, transponder, as a high-speed, large amount of real-time data transmission equipment, has become an important part of CTCS2 and above train control system. The transponder is widely used in railway field for its advantages of large amount of information, accurate positioning (error no more than 1 m), strong environmental adaptability, easy installation and long service life. Large-capacity point transponder system will become indispensable basic equipment in passenger dedicated lines and high-speed railways. At present, most transponders used in China are still abroad, domestic research and application are still in the initial stage, and domestic transponders can not be widely recognized. It is necessary to establish a test platform for transponders for further research and performance testing. For an environment. Transponders transmission system has become a key part of high-speed train operation—train operation control system.

5.6.2 Analysis of Uplink Signal of Transmitter

For a better understanding of transponders transmission system, we further execute a series of experiments to analyze static and dynamic characteristics of uplink signals. Static characteristics of the upstream signal refer to the waveform characteristics of the upstream signal transmitted by the ground transponder received by the vehicle antenna when the train and the ground transponder remain relatively static and the vehicle receiving antenna is located within the contact area of the transponder. Relatively static, the upstream signal is a FSK modulation signal with continuous phase, which is only a function of amplitude, frequency and phase, and its central frequency is 4.234 MHz + 200 kHz, modulation frequency offset 282.24 kHz + 5%, data rate 564.48 kbit/s + 2.5%. The modulated signal is BCH-coded transponder message with frequency of 3.651 MHz stands for logic 0 and frequency 4.516 MHz stands for logic 1; the dynamic characteristics of the up-link signal refer to the waveform characteristics of the up-link signal transmitted by the vehicle antenna when the train and the ground transponder are in relative motion. At this time, the amplitude of the up-link signal, especially the envelope curve of the signal received by the vehicle antenna at different train speeds, that is, the dynamic interaction time between the ground transponder and the vehicle antenna will change. However, different train speeds have little effect on signal frequency and phase characteristics. Based on the basic theory of electromagnetic field, this paper studies the antenna magnetic field distribution model of transponder transmission system under the condition of train operation, in order to obtain the dynamic characteristics of the up-link signal at different train speed.

Based on the above experimental analyses, we present working principles of uplink signal system. The function of transponder system is to transmit data between ground and vehicle equipment in a specific location. Vehicle equipment includes BTM antenna and BTM, ground equipment includes LEU (ground electronic unit) and active or passive transponder. Among them, passive transponder stores some fixed information, including line gradient information, fixed speed limit information, track circuit parameter information and train control system level conversion information; active transponder stores real-time information transmitted by LEU, including route information, temporary speed limit information and so on. The transponder is usually in a dormant state. Only when the train passes over the transponder, the transponder is activated by the high frequency electromagnetic energy transmitted by the vehicle-borne BTM antenna, and the stored message is modulated into an up-link signal in 2FSK mode and sent out at a certain frequency. After receiving the up-link signal, the vehicle-borne BTM antenna transmits it to the BTM module, filters, demodulates and decodes it to get the message information of the transponder, and transmits it to the vehicle-borne ATP for control. When the train is far away from the ground transponder, the transponder goes to sleep again.

5.6.3 Constructing the Simulation System on Transponder Transmission

Before explaining how to construct the simulation system, we need further understand the meaning of simulation technology. "Simulation is a model-based activity", which involves knowledge and experience in many disciplines and fields. The key to successful simulation research is to organize and implement various activities of simulation life cycle organically and coordinately. Here the "various activities" are "system modeling", "simulation modeling", "simulation experiments", and the elements associated with these activities are "system", "model" and "computer". Among them: the system is the object of study, the model is the abstraction of the system, and the simulation is to achieve the purpose of research through the experiment of the model. The relationship between elements and activities is shown in Fig. 5.5. Traditionally, the activity of "system modeling" belongs to the category of system identification technology, while simulation technology focuses on "simulation modeling", that is to say, to study its solving algorithm for different forms of system models, so that it can be realized on the computer. As for the activity of

Fig. 5.5 The relationship between elements and activities

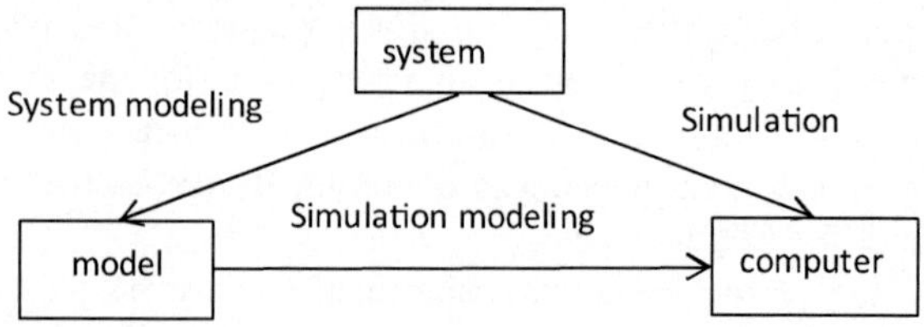

"simulation experiment", it usually pays attention to the verification of its "simulation program". As for the fundamental problem of how to compare the simulation experiment results with the actual system behavior, it lacks of methodological research.

Finally, we will theoretically discuss how to establish the simulation system. Traditionally, a system model is established after experimental identification. In recent ten years, the technology of system identification has developed rapidly. In identification methods, there are time domain method, frequency domain method, correlation analysis method, least squares method, etc. In technical means, there are systematic identification design, system model structure identification, system model parameter identification, system model verification, etc. in simulation modeling. In addition to adapting to the development of computer software and hardware environment and constantly researching new algorithms and developing new software, modern simulation technology uses model and experiment separation technology, namely model data driven. The model is divided into parameter model and parameter value in order to improve the flexibility and efficiency of simulation; simulation experiments. Modern simulation technology distinguishes the experimental framework from the simulation operation control. The experimental frame is used to define conditions, including model parameters, input variables, observation variables, initial conditions and output instructions. In this way, when different forms of output are needed, it is not necessary to modify the simulation model, or even to re-simulate the operation.

To implement computer simulation for a system or object, we must first grasp the basic characteristics of the object, grasp the main factors, introduce the necessary parameters, put forward reasonable assumptions, Abstract scientifically, analyze the relationship between the parameters, and select appropriate mathematical tools, and then establish the corresponding mathematical model on this basis. The model is the simulation of the attributes of the object studied. It should not only reflect the basic characteristics of the objective object and the main properties of our interest, but also be easy to process on the computer. Therefore, it is often necessary to modify the above steps repeatedly, which is usually referred to as secondary modeling. Finally, a computer program is compiled, and the relevant conditions are changed to analyze and process the calculation results, from which the basic characteristics of the objective object are studied.

Computer simulation is a new technology with strong flavor of the times. It is widely used in modern scientific research, engineering design, business decision-making, national defense construction and many other fields. In the field of modern scientific research, the development of science and technology is advancing by leaps and boundaries of new disciplines, interdisciplinary disciplines and frontier disciplines. How to prove the correctness of these new ideas and verify their experimental feasibility, and how to solve some key problems in frontier science, these can be verified or made a breakthrough by means of computer simulation. When scientists are faced with a scientific and technological problem that needs to be solved urgently, whether the existing solutions conform to reality, whether new problems will arise from them, and how to grasp the related complex factors. Computer simulation can bring great convenience. It frees scientists and technicians from the complicated experimental equipment in the early stage of scientific research. It not only saves manpower and material resources, but also improves work efficiency. Therefore, computer simulation has become an extremely important means of modern scientific research.

This section was explicitly reorganized from [72]. Computer simulation plays an extremely unique and important role, and has remarkable advantages that other means can not match. In a word, computer simulation technology, with its vigorous vitality, appears in all fields of today's society. It has profoundly changed people's way of thinking, working and living, and brought good social and economic benefits and in the future will also contribute to evolution of the machine brain.

References

1. Wang W. F., Deng X. Y., Ding L., et al. Pheromone Accumulation and Iteration, Brain-inspired Intelligence and Visual Perception. Springer, 2019: 41–58.
2. Wang W. F., Deng X. Y., Ding L., et al. The Vision Brain Hypothesis, Brain-inspired Intelligence and Visual Perception. Springer, 2019: 17–31.
3. Dorigo M. "Optimization, Learning and Natural Algorithms". Ph.D. Thesis, Department of Electronics, Politecnico diMilano, Italy, 1992.
4. Gambardella L M, Dorigo M. "Ant Q: a reinforcement learning approach to the traveling salesman problem." Proceedings of the 12th International Conference on Machine Learning. 1995, 252–260.
5. Dorigo M, Gambardella L M. "Ant colony system: a cooperative learning approach to the traveling salesman problem". IEEE Transactions on Evolutionary Computation. 1997 1(1):53-66.
6. Maniezzo V, Colorni A, Dorigo M. "The ant system applied to the quadratic assignment problem". Technical report IRIDIA/94–28, IRIDIA, Université Libre de Bruxelles, Belgium, 1994.
7. Gao Q L, Luo X, Yang S Z. "Stigmergic cooperation mechanism for shop floor control system". International Journal of Advanced Manufacturing Technology. 2005, 25(7-8):743-753.
8. Ali Onder Bozdogan, Murat Efe. "Improved assignment with ant colony optimization for multi-target tracking". Expert Systems with Applications. 38, 2011, 9172–9178.
9. W. Xiang, H.P. Lee. "Ant colony intelligence in multi-agent dynamic manufacturing scheduling". Engineering Applications of Artificial Intelligence. 21, 2008, 73-85.
10. John E. Bella, Patrick R. McMullen. "Ant colony optimization techniques for the vehicle routing problem". Advanced Engineering Informatics. 18, 2004, 41-48.
11. Wang L, Wu Q D. "Linear system parameters identification based on Ant system algorithm". Proceedings of the IEEE Conference on Control Application. 2001, 401–406.

12. Blum C, Dorigo M. The hyper-cube framework for ant colony optimization. IEEE Transactions on Systems, Man and Cybernetics—Part B; to appear. Also available as Technical Report TR/IRIDIA/2003–03, IRIDIA, Université Libre de Bruxelles, Belgium; 2003.
13. Colorni A, Dorigo M, Maniezzo V. Distributed optimization by ant colonies. Proceedings of the 1st European Conference on Artificial Life. Paris, 1991, 134–142.
14. Stützle T, Hoos H. H. The Max-Min Ant System and local search for the traveling salesman problem. In T. Back, Z. Michalewicz, & X. Yao (Eds.), Proceedings of the 1997 IEEE International Conference on Evolutionary Computation (ICEC'97) (pp 309–314).
15. Bullnheimer B, Hartl R F, Strauss. A new rank-based version of the ant system: a computational study. Central European Journal of Operations Research and Economics, 1999, 7(1): 25–38.
16. HUANG Lan, WANG Kangping, ZHOU Chunguang et al. Hybrid Approach Based on Ant Algorithm for Solving Traveling Salesman Problem. JOURNAL OF JILIN UNIVERSITY (SCIENCE EDITION), 2002, 40(4): 369–373.
17. HAO Jin, SHI Libao, ZHOU Jiaqi. An Ant System Algorithm with Random Perturbation Behavior for Complex TSP Problem. Systems engineering-theory & practice, 2002, 9: 88-91.
18. Chi-Bin Cheng, Chun-Pin Mao. A modfied ant colony system for solving the travelling salesman problem with time windows. Mathematical and Computer Modelling, 2007, 46: 1225–1235.
19. Jinhui Yang, Xiaohu Shi, Maurizio Marchese, et al. An ant colony optimization method for generalized TSP problem. Progress in Natural Science, 2008, 18: 1417–1422.
20. C. Blum, M. Dorigo. The hyper-cube framework for ant colony optimization. IEEE Transactions on Systems, Man and Cybernetics-Part B. Also available as Technical Report TR/IRIDIA/2003–03, IRIDIA, Universit Libre de Bruxelles, Belgium; 2003.
21. Gambardella L M, Dorigo M. Ant-Q: a reinforcement learning approach to the traveling salesman problem. Proceedings of the 12th International Conference on Machine Learning. 1995:252–260.
22. Xiangyang Deng, Limin Zhang, Hongwen Lin, et al. Pheromone Mark Ant Colony Optimization with a Hybrid Node-based Pheromone Update Strategy. Neurocomputing, In Press, https://doi.org/10.1016/j.neucom.2012.12.084.
23. J. Q. Geng, L. P. Weng, S. H. Liu. An improved ant colony optimization algorithm for nonlinear resource-leveling problems. Computers and Mathematics with Applications, 61, 2300-2305, 2011.
24. Colorni A, Dorigo M, Maniezzo V, et al. "Distributed optimization by ant colonies". Proceedings of the 1st European Conference on Artificial Life. Paris, 1991, 134–142.
25. Dorigo M, et al. "Positive feedback as a search strategy". Technical report 91–016, Department of Electronics, Politecnico diMilano, Italy, 1991.
26. Gao S, Zhong J, MO S J. Research on ant colony algorithm for continuous optimization problem. Microcomputer Development, 2003, 13(11):12–13.
27. Chen X, Yuan Y. Novel ant colony optimization algorithm for robot path planning. Systems Engineering and Electronics, 2008, 30(5):952-955.
28. Abe Y, Shikann M, Fokuda T, et al. "Vision based navigation system by variable template matching for autonomous mobile robot". Proceedings of the IEEE International Conference on Robotics & Automation. Leaven, 1998, 952–957.
29. Weerayuth N, Chaiyaratana N. "Closed-loop time-optimal path planning using a multi-objective diversity control oriented genetic algorithm". IEEE International Conference on Man and Cybernetics. 2002, 7 pp. vol 6.
30. S. Goss, S. A., J.L. Deneubourg et al. "Self-organized shortcuts in the argentine ant". Naturwissenschaften. 1989, 76: 579–581.
31. Dorigo M, Stützle T.: Ant Colony Optimization. The MIT Press, Cambridge, Massachusetts London, England. 2004, 25-221.
32. J.-L. Deneubourg, S. A., S. Goss, et. al. "The Self-Organizing Exploratory Pattern of the Argentine Ant". Journal of Insect Behavior. 1990, 3(2): 159–168.
33. Hungr N, Bricault I, Cinquin P, et al. Design and Validation of a CT- and MRI-Guided Robot for Percutaneous Needle Procedures [J]. IEEE Transactions on Robotics, 2016, 32(4):973-987.

34. Chen A I, Balter M L, Maguire T J. Developing the World's First Portable Medical Robot for Autonomous Venipuncture[J]. IEEE Robotics & Automation Magazine, 2016, 23(1):10-11.
35. Vuurmans T, Er L, Sirker A, et al. Long-term patient and kidney survival after coronary artery bypass grafting, percutaneous coronary intervention, or medical therapy for patients with chronic kidney disease: a propensity-matched cohort study [J]. Coronary Artery Disease, 2018, 29(1):8-16.
36. Li F, Tai Y, Shi J, et al. Dynamic Force Modeling for Robot-Assisted Percutaneous Operation Using Intraoperative Data [C]. International Conference on Mechatronics and Intelligent Robotics. Springer, 2017, 554–560.
37. Su H, Shang W, Li G, et al. An MRI-Guided Telesurgery System Using a Fabry-Perot Interferometry Force Sensor and a Pneumatic Haptic Device [J]. Annals of Biomedical Engineering, 2017(1):1-12.
38. Alaid A, Von E K, Smoll N R, et al. Robot guidance for percutaneous minimally invasive placement of pedicle screws for pyogenic spondylodiscitis is associated with lower rates of wound breakdown compared to conventional fluoroscopy-guided instrumentation [J]. Neurosurgical Review, 2017(Suppl 1):1–8.
39. Dagnino G, Georgilas I, Morad S, et al. Image-Guided Surgical Robotic System for Percutaneous Reduction of Joint Fractures [J]. Annals of Biomedical Engineering, 2017(6):1-15.
40. Smitson C, Ang L, Reeves R, et al. TCT-124 Safety and Feasibility of a Novel, Second Generation Robotic-Assisted System for Percutaneous Coronary Intervention: First-in-Human Report [J]. Journal of the American College of Cardiology, 2017, 70(18): B55-B56.
41. Grassano Y, Cornelis F, Grenier N, et al. Minimally invasive conservative treatment of localized renal tumors: A single center experience on percutaneous ablations and robot-assisted partial nephrectomy [J]. The journal of urology, 2017, 197(4S): 1-1.
42. Chu C M, Yu S. Robot-assisted navigation system for CT-guided percutaneous lung tumour procedures: our initial experience in Hong Kong [J]. Cancer Imaging, 2014, 14(1):1-1.
43. Giulianotti P C, Quadri P, Durgam S, et al. Reconstruction/Repair of Iatrogenic Biliary Injuries: Is the Robot Offering a New Option? Short Clinical Report [J]. 2018, 267(1): e7-e9.
44. De J P, Huyse F J, Slaets J P, et al. Care complexity in the general hospital: results from a European study [J]. Psychosomatics, 2001, 42(3):204-212.
45. Tondut L, Peyronnet B, Bernhard J C, et al. Impact of hospital volume and surgeon volume on robot-assisted partial nephrectomy results: A multicenter study [J]. European Urology Supplements, 2017, 16(3):e1130-e1134.
46. Brandmeir N J, Sather M D. A Technical Report of Robot-Assisted Stereotactic Percutaneous Rhizotomy [J]. Pain Medicine, 2017, 18 (12):2512-2514.
47. Spataro R, Chella A, Allison B, et al. Reaching and Grasping a Glass of Water by Locked-In ALS Patients through a BCI- Controlled Humanoid Robot [J]. Frontiers in Human Neuroscience, 2017, 11(3): 1-10.
48. Chhor V, Follin A, Joachim J, et al. Risk factors of percutaneous cannulation failure by intensivists for veno-arterial extracorporeal life support for refractory cardiac arrest [J]. Intensive Care Medicine, 2017, 43(11):1742-1744.
49. Trinh Q D, Sammon J, Sukumar S, et al. 14 Long-term follow-up of patients undergoing percutaneous suprapubic tube drainage following robot assisted radical prostatectomy [J]. European Urology Supplements, 2011, 10(8):536-536.
50. H. J. Cai, Tianqi Cai, Wenwei Zhang and Kai Wang. The beginning and evolution of consciousness. Advances in Intelligent Systems Research, Vol. 133, pp. 219–222, 2016.
51. Penrose R. The emperor's new mind. Oxford: Oxford University Press, 1989.
52. Minsky M. The emotion machine. Simon & Schuster. 2007.
53. Donald D. Hoffman, Chetan Prakash. Objects of machine minds. Frontiers in Psychology. 5(5):606-626, 2014.
54. Portmann A. A zoologist looks at humankind. New York: Columbia University Press, 1990.
55. Kay P, McDaniel C K. The linguistic significance of the meaning of basic color term. Language, 1978, 54(5): 610–646.

56. Cai H J, Tian X. Chinese economic miracles under the protection of the cognitive membrane. Conference on Web Based Business Management, Shanghai, 2012, 606–610.
57. Cai H J. The historical context of the rise of china and the entry point of the transformation of the development pattern. Emergence and Transfer of Wealth, 2012, 2: 1–6.
58. Cai H J, Cai T Q. Language acquisition and language evolution associated with self-assertiveness demands. Advances in Social and Behavioral Sciences, 2013, 2: 261–264.
59. Schwarz S. Exploring the physical layer frontiers of cellular uplink [J]. Eurasip Journal on Wireless Communications & Networking, 2016, 2016(1):118.
60. Rabén H, Borg J, Johansson J. A CMOS front-end for RFID transponders using multiple coil antennas [J]. Analog Integrated Circuits & Signal Processing, 2015, 83(2):149-159.
61. Hayami T, Iramina K, Chen X. Computer Simulation of Nerve Conduction Study of a Sural Nerve to Evaluate Human Peripheral Nervous System[C]// International Conference on the Development of Biomedical Engineering in Vietnam. 2017.
62. Thafasal Ijyas V P, Sameer S M. A reduced-complexity bat algorithm for joint CFO estimation in OFDMA uplink[C]// IEEE International Conference on Signal Processing. 2015.
63. Wang W Q. Transponder-Aided Joint Calibration and Synchronization Compensation for Distributed Radar Systems [J]. Plos One, 2015, 10(3):e0119174.
64. Ding Z, Schober R, Poor H V. A General MIMO Framework for NOMA Downlink and Uplink Transmission Based on Signal Alignment [J]. IEEE Transactions on Wireless Communications, 2016, 15(6):4438-4454.
65. Farhang A, Marchetti N, Doyle L E, et al. Low Complexity CFO Compensation in Uplink OFDMA Systems With Receiver Windowing [J]. IEEE Transactions on Signal Processing, 2015, 63(10):2546-2558.
66. Keyhani M. Computer Simulation Studies of the Entrepreneurial Market Process[M]// Complexity in Entrepreneurship, Innovation and Technology Research. 2016.
67. Zöscher L, Grosinger J, Spreitzer R, et al. Concept for a security aware automatic fare collection system using HF/UHF dual band RFID transponders[C]// Solid State Device Research Conference. 2015.
68. Brunner B, Pfeiffer F, Finkenzeller K, et al. Impact of GSM Interference on passive UHF RFIDs[C]// Smart Systech; European Conference on Smart Objects. 2015.
69. Terada T, Fukuda H, Kuroda T. Transponder Array System with Universal On-Sheet Reference Scheme for Wireless Mobile Sensor Networks without Battery or Oscillator[C]// Solid-state Circuits Conference Digest of Technical Papers. 2015.
70. Iyer S, Singh S P. Spectral and power efficiency investigation in single- and multi-line-rate optical wavelength division multiplexed (WDM) networks [J]. Photonic Network Communications, 2017, 33(1):39-51.
71. Wang D, Min Z, Jiao Z, et al. Evaluations of transponder bank size and network performance for CD-ROADMs[C]// International Conference on Optical Communications & Networks. 2015.
72. Ning Z, Ying H. Transponder Uplink Signal Computer Simulation System [C]. International Conference on Frontier Computing, 2019.

Chapter 6
Interdisciplinary Evolution of the Machine Brain

Abstract This chapter finally presents a characterization of interdisciplinary evolution of the machine brain. Perspective schemes for rebuilding a real vision brain in the future are analyzed, along with the major principles to construct the machine brain, are presented, which include memory, thinking, imagination, feeling, speaking and other aspects associated with machine vision, machine touch and machine minds. This explicitly developed the theoretical framework of brain-inspired intelligence from the vision brain hypothesis, the vision-minds hypothesis and the skin brain hypothesis. Based on Chaps. 2–5, development of machine intelligence during the past decades have experienced three stages—machine computation, learning and understanding. Machine leaning includes data mining. Environmental sensing helps to acquire the data. Pattern analysis and scene understanding are significant parts of machine understanding. Scientists have taught machine how to collect and treat data and discover knowledge from the data. Evolution of machine brain will experience another two stages—machine meta-learning (learning to learn) and self-directed development (improving the capability of machine brain utilizing the learned knowledge). There are still great challenges in realization of the dream.

6.1 Background

In all external information obtained by human brain, vision information occupies a large proportion [1–4], especially for primates. For a normal person, vision information accounts for more than 70% of all perceptive information [3]. The dorsal pathway includes a series of brain regions (V1/V2-MT-MST/VIP) of the occipital lobe to the parietal lobe, which mainly deal with motion and depth related visual perception. The ventral occipital regions of the brain including a series of leaf Daonie leaves (V1/V2-V4-TEO/IT), mainly dealing with the shape and color information. Along with the hierarchical transmission of image information along the visual pathway, the information extracted from functional brain area from visual images and changes is also from simple to complex, from concrete to abstract. For example, V1 neurons can distinguish small rod orientation, space position and direction of movement;

W. Wang et al., *Interdisciplinary Evolution of the Machine Brain*,
Research on Intelligent Manufacturing,
https://doi.org/10.1007/978-981-33-4244-6_6

consistency of MT neurons can distinguish between multiple object moving direction, and their relative position relationship; to advanced visual cortex MST and VIP region, where neurons can even through the comprehensive analysis of the rotation motion of the object in all feelings and the eyeball, accurately infer the direction of movement of the viewer. It should be explained that the dorsal and ventral pathways are only a rough classification of the brain regions based on the main functions of visual information processing. In fact, the nervous system is a network structure. There are bidirectional projections not only in all pathways, but also in all two brain regions.

Visual information processing is the core function of the human brain, and the area of the cerebral cortex about 1/4 is involved in this work [7–9]. At present, the brain's processing of visual information follows 3 principles: one is distributed, that is, different functional brain areas perform their duties, such as object orientation, motion direction, relative depth, color and shape information, etc., which are handled by different brain regions [6]. The two level processing, information processing, and has the initial path composed of higher brain areas, primary cortical resolution contrast, brightness, color, orientation and direction of motion of a single object and multiple objects, intermediate cortical discrimination between moving objects in the scene relation, spatial layout and surface characteristics, to distinguish the foreground and background so, can the higher cortical complex environment of the object recognition, with other interference factors, perception of visual perception information to exclude the impact of the stability of lead in different parts of the body to interact with the environment behavior [10]. The three is the networking process, that is, there is a wide range of interaction and projection between each functional area of brain visual information processing, similar to subordinate reporting to superiors, superiors to subordinate instructions, and colleagues to coordinate with each other. Taking into account the access to external image and its change is consistent, because the information processing limits of the visual system (for example, each neuron can only see a small piece of vision), and how to realize the distributed analysis and processing on the image, the network organization is perhaps people can form a necessary guarantee of stability, visual perception unified [11]. In addition, the visual system is only part of the cerebral nervous system, which causes the process of information processing inevitably affected by the big environment, and vice versa [12]. These factors include emotional state, experience preference, and attention target. For example, too much attention to a thing will often turn a blind eye to the surrounding scenery, and the more familiar with something, the easier it is to identify it from the complex background.

In addition to the visual system, People are always to utilize auditory system and tactile system etc. to obtain more external information [13–18]. In most cases, the mentioned sensory systems need to cooperate with each other so that to obtain more accurate cognition of the surroundings [19–21].

Cognition is the process of acquiring knowledge or applying knowledge, or process of information processing, which is the most basic psychological process of human being. It includes feeling, perception, memory, thinking, imagination, and language. The information input by human brain is processed by mental processing, and then transformed into internal mental activity, which controls human behavior. This process is the process of information processing, that is, cognitive process. The cognitive ability of human beings is closely related to the process of human cognition, and it can be said that cognition is a product of the process of human cognition. Generally speaking, people's perception of objective things (feeling, perception), thinking (imagination, association, thinking) are all cognitive activities. The process of cognition is the process of subjectively objectification, that is, the subjective reflection of the objective and the objective expression in the subjective [22].

Cognition consists of feeling, perception, memory, thinking, imagination, and language. In particular, the process of acquiring knowledge or applying knowledge begins with feeling and perception. Feeling is the understanding of the individual attributes and characteristics of things, such as color, shade, tone, smell, fine, soft and hard. And perception is the cognition of the whole and its relations and relations, such as seeing a red flag, hearing a noisy voice and touching a light sweater. People's knowledge and experience gained by perception do not disappear immediately after the stimulus stops, and it is retained in people's minds and can be reproduced when needed. The psychological process of accumulating and preserving individual experience is called memory. People can not only direct perception, individual specific things, surface contact and relationship to understand things, but also use the acquired knowledge and experience to the indirect and general understanding of the things, revealing the essence of things and the inherent relationship between the formation of the concept of things, for reasoning and judgment, solve all sorts of face this is the problem of thinking [23].

People can also use language to communicate their own thinking activities, the results of cognition activities and others' experience, which is language activity. People also have imaginary activities, which are carried out by the specific image that is preserved in the mind. "Cognitive process". (1) the traditional cognitive process refers to the psychological process of the human brain reflecting the characteristics of objective things and their relationships in the forms of perception, memory and thinking. (2) the Piaget School of language refers to the process of assimilating and adapting to the stimulus through the original cognitive structure (schema). This is the view of structuralist cognitive psychology. (3) the use of cognitive psychology in information processing refers to the process of individual acceptance, coding, storage, extraction, and use of information. It usually consists of 4 components: perception system (receiving information), memory system (information coding, storage and extraction), control system (supervision execution decision), reaction system (control information output) and so on [24].

It was recognized that there are 3 basic processes in human cognition: (1) problem solving. The method of elicitation, means—purpose analysis and planning process are used. (2) pattern recognition ability. In order to establish the pattern of things, one must recognize the relationship between the elements. Such as an equivalent relationship, a continuous relationship, etc. According to the relationship between elements, it can form a pattern. (3) learning. Learning is to get information and store it so that it can be used later. There are different forms of learning, such as learning, reading, understanding, and example learning [25–27].

As a whole, cognitive process is the process of information processing of individual cognitive activities. Cognitive psychology regards cognitive process as a system of information processing based on a series of continuous cognitive operation stages, such as information acquisition, coding (Encoding & Coding), storage, extraction and use. The acquisition of information is to accept the stimulus information that directly acts on the senses. The function of feeling is to get information. The encoding of information is to convert one form of information into another form of information in order to facilitate the storage, extraction and use of information. Individuals have corresponding information encoding methods in cognitive activities such as perception, imagery, imagination, memory and thinking. The storage of information is the retention of information in the brain. In memory activities, the storage of information has many forms. The extraction of information is based on a certain clue to find the information needed from memory and take it out. The use of information is to use the extracted information to process the new information. In the cognitive process, by encoding information, the characteristics of external objects can be transformed into specific information, such as specific images, semantics or propositions, and then stored in the brain. These concrete images, semantics and propositions are the manifestations of the characteristics of the external object in the individual psychology, and the reflection of the objective reality in the brain. The specific images, semantics or propositions that reflect the characteristics of objective objects in the brain are called psychological representations of external objects, or Representation. Generally, "characterization" also refers to the process of processing the information of the external object in a certain form in the brain [28].

A policy or method of decision. It is the process of making ideas and making decisions for all kinds of events. It is a complicated process of thinking and operation, which is the process of collecting and processing information, making judgment and drawing conclusions. "Hanfeizi" indignation: "wise decision process for fools, the Magi found norikata intelligent people shy people on the contrary, the main." Decision making is not only an important ideological activity, but also an important part of human behavior. The basic structure of the decision can be roughly divided into three stages according to the difference of time and function. One is to evaluate the utility of the options and to make a choice. The two is the implementation of the last stage of choice. In this process, a series of process related to execution need to be completed, such as selecting the order of different actions, prohibiting other competitive behaviors and selecting suitable action time. The three is the experience of decision, the difference between the expected and actual results mainly from experience, has the important significance of this difference on the adjustment of the assignment decision options. It can be seen that decision making is an iterative process, and the decision of the current stage is not only affected by the previous stage, but also by the later stage [24].

6.2 Practical Multi-modules Integration

Human intelligence behavior is carried out through a variety of sensory and sensory channels, such as the combination of vision and hearing, the combination of vision and balance, and audio-visual fusion is the most common way of integration. Vision and hearing are the main incoming passages of human information, and also the basis for the formation of cognition. Through the visual system, we can perceive the size, shape, color, shade, motion, distance, and so on of the external objects. Through auditory sense, we can feel the stimulation of voice, recognize the meaning of voice, and talk, exchange ideas, music appreciation, find signals and avoid danger on this basis. Audio-visual function is closely related to many advanced functions of the brain. Visual and auditory fusion is an important approach to the processing mechanism of human brain information [30].

6.2.1 *Scheme for Integration*

When faced with complex environment in daily life, various sensory systems such as vision, hearing and so on are always involved in a specific interaction so as to meet the needs of living and living of organisms. First, each sensory system can provide a “defect fragment”. For example, the visual system generally provides space information well, and can not find the location of an object, an object outside the field of vision, or a dark object in the dark. The auditory system can fill these gaps. In noisy environment, visual information has a very significant compensation effect on auditory information, which can greatly improve the correct recognition rate of people’s perception of language information. The natural supplement of audiovisual space information provides significant survival significance. It is beneficial for humans to walk through the crowded roads, animals to hunt or avoid being arrested. Secondly, when both visual and auditory information is provided for the same object or event, joint signals from each feature can improve the accuracy of perception. The experimental results show that when the sound and optical signals to the same spatial orientation, the correct rate of animal toward the reaction to improve the reaction time shortened, identification error is significantly reduced and given as a combination of stimulation, spatially separated, the animal towards the right rate, reaction time is prolonged and the discrimination error increased significantly. Therefore, the sound and visual signals from time and space can improve the discrimination ability of animals and humans. On the contrary, the two signals separated from space and time will reduce the ability of animals and humans to distinguish objects. As the processing of the nervous system changes, permanent differences in visual and auditory information can lead to permanent changes in the two sensory transduction pathways [31, 32].

Audiovisual interaction. Vision and hearing use their respective spatial and temporal advantages to provide different clues for information cognition. However, how to integrate visual and auditory functions when recognizing the same object or event? Zhao Lina and other EEG experiments under visual and auditory stimulation mode showed that the synchronization index between the occipital and temporal lobe was significantly greater than that between the occipital and parietal lobe. It can be seen that in the simultaneous use of visual and auditory modes, both visual and auditory regions automatically synchronize neural activity, which promotes the integration of audio-visual information in the brain. Behavioural studies have shown that visual clues have a strong influence on auditory spatial perception. Further research by the Kjludsen team found that the barn owl has a fixed projection pathway from the visual cortex to the central and peripheral nuclei of the inferior colliculus. The visual cortex can provide guidance clues for the expression of auditory localization in the inferior colliculus, so that special visual cues can be linked with spatial location and form auditory spatial perception. The results of Busse and other studies reveal that auditory cortex's response to sound is influenced by visual attention. Zhang Wei and other audiovisual evoked potentials showed that the auditory brainstem response (ABR) of auditory evoked potential was compared with that of ABR alone, and the latency of partial auditory wave was longer than that of the latter [33].

It is suggested that visual information has time guidance on auditory information in the process of audio-visual information integration (influence latency). Gelde studies show that in the human brain, there are many sensory neurons. These neurons can make the specific sensory regions of the brain react with various stimuli at the same time. They exist in many brain regions, most of which are the apical cortex, the deep upper colliculus and some cortical parts, such as the anterolateral trench and the external superior temporal trench. Electrophysiological studies show that from space and time consistent visual and auditory information converge to the same number of sensory neurons, also fall into the excitatory receptive field, can induce neuronal electrical activity produced strong. On the contrary, when any one of the visual and auditory stimuli is outside the excitement of the field, it inhibits the response of the multisensory neurons. In short, in the process of integration of visual and audio information, visual cortex nerve cell synchronous discharge, visual information of time and space to guide the auditory information, and there are many sensory neurons of visual and auditory information exchange, so that the brain activities of balance and coordination in function, and to maintain continuity of cognition and behavior [34].

6.2.2 Vision and Auditory Integration

Human intelligence behavior is carried out through a variety of sensory and sensory channels, such as the combination of vision and hearing, the combination of vision and balance, and audio-visual fusion is the most common way of integration. Vision and hearing are the main incoming passages of human information, and also the basis for the formation of cognition. Through the visual system, we can perceive the size, shape, color, shade, motion, distance, and so on of the external objects.

Through auditory sense, we can feel the stimulation of voice, recognize the meaning of voice, and talk, exchange ideas, music appreciation, find signals and avoid danger on this basis. Audio-visual function is closely related to many advanced functions of the brain. Visual and auditory fusion is an important approach to the processing mechanism of human brain information [35].

When faced with complex environment in daily life, various sensory systems such as vision, hearing and so on are always involved in a specific interaction so as to meet the needs of living and living of organisms. First, each sensory system can provide a "defect fragment". For example, the visual system generally provides space information well, and can not find the location of an object, an object outside the field of vision, or a dark object in the dark. The auditory system can fill these gaps. In noisy environment, visual information has a very significant compensation effect on auditory information, which can greatly improve the correct recognition rate of people's perception of language information. The natural supplement of audiovisual space information provides significant survival significance. It is beneficial for humans to walk through the crowded roads, animals to hunt or avoid being arrested. Secondly, when both visual and auditory information is provided for the same object or event, joint signals from each feature can improve the accuracy of perception. The experimental results show that when the sound and optical signals to the same spatial orientation, the correct rate of animal toward the reaction to improve the reaction time shortened, identification error is significantly reduced and given as a combination of stimulation, spatially separated, the animal towards the right rate, reaction time is prolonged and the discrimination error significantly increased [5]. Therefore, the sound and visual signals from time and space can improve the discrimination ability of animals and humans. On the contrary, the two signals separated from space and time will reduce the ability of animals and humans to distinguish objects. As the processing of the nervous system changes, permanent differences in visual and auditory information can lead to permanent changes in the two sensory transduction pathways. Audiovisual interaction. Vision and hearing use their respective spatial and temporal advantages to provide different clues for information cognition. However, how to integrate visual and auditory functions when recognizing the same object or event? Zhao Lina and other EEG experiments under visual and auditory stimulation mode showed that the synchronization index between the occipital and temporal lobe was significantly greater than that between the occipital and parietal lobe. In the simultaneous use of visual and auditory modes, both visual and auditory regions automatically synchronize neural activity, which promotes the integration of audio-visual information in the brain. Behavioral studies have shown that visual clues have a strong influence on auditory spatial perception. Further research has found that there is a fixed projection pathway from the visual cortex to the central and peripheral nuclei of the inferior colliculus in the barn owl. The visual cortex can provide guidance clues for the expression of auditory localization in the inferior colliculus, so that special visual cues can be linked with spatial location and form auditory spatial perception. The results of Busse and other studies reveal that auditory cortex's response to sound is influenced by visual attention. Audio-visual evoked potentials in Zhang Wei and so on showed that the auditory brainstem response (ABR) of auditory evoked potential was compared with that of ABR alone, and the latency of partial auditory wave was longer than that of the latter [29–35].

It is suggested that visual information has time guidance on auditory information in the process of audio-visual information integration (influence latency). Some studies show that in the human brain, there are many sensory neurons. These neurons can make the specific sensory regions of the brain react with various stimuli at the same time. They exist in many brain regions, most of which are the apical cortex, the deep upper colliculus and some cortical parts, such as the anterolateral trench and the external superior temporal trench. Electrophysiological studies show that from space and time consistent visual and auditory information converge to the same number of sensory neurons, also fall into the excitatory receptive field, can induce neuronal electrical activity produced strong. On the contrary, when any one of the visual and auditory stimuli is outside the excitement of the field, it inhibits the response of the multisensory neurons. In short, in the process of integration of visual and audio information, visual cortex nerve cell synchronous discharge, visual information of time and space to guide the auditory information, and there are many sensory neurons of visual and auditory information exchange, so that the brain activities of balance and coordination in function, and to maintain continuity of cognition and behavior [36–38].

6.3 Practical Multi-model Fusion

Multimodal fusion is an integral part of information fusion. According to the different data sources, multi-modal fusion can be divided into sensor level fusion (also called data level fusion), feature level fusion, knowledge level fusion and decision level fusion.

Next, we will take this as a classification basis and summarize the methods and main ideas used by each level of integration.

6.3.1 Sensor Layer Fusion

Sensor layer fusion aims to integrate data collected by each sensor and form a new set of data and send it to the subsequent feature extraction module. In the field of image processing, this is also the image fusion or pixel fusion. Because the signals and data collected by sensors are the most original, they don't lose much information because of feature extraction or template matching. These data contain a large amount of credible information. Compared with other algorithms, many reliable data become the advantage of sensor level fusion algorithm, and theoretically speaking, the advantage of sensor layer algorithm should achieve the best fusion effect. In the general sensor level fusion framework, the system can acquire more complete output that can describe the data source characteristics by collecting multiple features of the same data source and integrate these data to restore the corresponding object feature information. For example, in the fingerprint recognition sensor fusion algorithm, the smaller sensor used to collect fingerprints, a single sampling was difficult to obtain all the fingerprint texture, so the system usually adopts the scheme of fingerprint image acquisition, repeatedly registered the same finger, it can obtain the fingerprint information of the finger under all kinds of gestures, then the fingerprint images fusion into a contains the fingerprint information of all edge structure complete image. It is proposed a method using finger rolling to collect complete fingerprint information, however, this method requires the system to use a large area of sensor, so it has not been widely used. The researchers called the fusion method a splicing fusion and played a very good role in the actual hardware system [39].

6.3.2 Feature Layer Fusion

After collecting the original multi-modal signal data, these data will pass through their respective modal specific feature extraction methods to generate corresponding feature templates, and feature layer fusion is the process of updating or splicing these feature templates. According to the different fusion targets, the feature layer fusion method can be divided into two problems: template upgrading or template improvement, and heterogeneous feature fusion. Template upgrading or template improvement refers to the process of fusion that will be extracted from the same biological mode to the template by the same feature extraction method. Among them, template update is to update information in template based on new template information. A simple way to update templates is to generate new templates from average values of new and old two templates. The method of seed delivery is often applied to sometime varying biological signs, such as face, etc., and it can also be used to improve some permanent changes of biological modes, such as palmprint with scar marks. Template improvement is based on comparing old and new templates to remove redundant features and increasing the feature dimension of templates to enrich template information and perfect template details [40].

Heterogeneous feature fusion is to stitching together the feature vectors from different biological modalities, and to generate a feature vector which is easy to compare and has strong distinguishing ability through feature normalization and feature selection. Before feature splicing, feature normalization is often needed to avoid a series of problems caused by the different feature space of heterogeneous features. For example, analysis, there are two groups of feature vectors were extracted from different biological mode, the characteristics of each dimension respectively in range of [0, 1] and [0, 10]. Since the eigenvalues of the latter are generally larger than the former, when calculating the distance between the two sets of feature vectors, the distance values obtained, for example, the distance measured by Euclidean distance, are often dominated by the latter, while the former's contribution to distance measurement is much weaker. This will directly lead to the improvement of the accuracy rate after fusion because of the inefficient use of the feature meaning, which will also lose the significance and advantage of multi-modal fusion. Therefore, before the two sets of feature splicing, we also need to normalize the features. There are many ways to characterize normalization. Among them, the minimum maximum method is one of the simplest and effective methods. Besides, there are many similar normalization methods, such as the Absolute Median deviation method and so on [41].

It is worth noting that many fractional normalization methods in fractional level fusion are interworking with feature normalization methods, which will be discussed below in the next level. These normalization methods have their own merits and demerits, their emphasis is different, and the applicable situations vary widely. What normalization methods do we need to use specifically depends on the specific environment. After the normalization of the features, the heterogeneous features can be spliced together. However, due to the characteristics of simple splicing tend to generate a set of features very high dimensions, Yu will directly lead to the curse of dimensionality. At the same time, the feature space in such spliced, often contain many redundant features, and treatment of redundant features will bring additional repeated calculation, not only a waste of computing resources, but also increased the consumption of the fusion process of time. And, due to noise and other factors, some of the characteristics can also have a negative impact on the final recognition rate. Therefore, the feature selection method is usually used to overcome the above problems, and a set of most discriminative feature subsets is obtained. Kumar had the feature selection in biometrics is discussed in detail feature selection problem existing feature selection methods can be applied to multiple modality, such as forward selection method, backward selection method, floating forward selection method, floating backward selection method, Boosting method, the maximum correlation, minimum redundancy method, sparse method, and linear programming method. There are also ways to reduce the feature dimension by transforming the features, such as the principal canonical correspondence analysis (CCA), the component analysis (PCA) and the linear discriminant analysis (LDA), etc., through W transformation. Ross has used PCA and LDA for feature selection and has achieved good results in the fusion of face and palm [42–44].

6.3.3 Knowledge Layer Fusion

In multi modal fractional layer, information fusion is the comparison of scores through specific feature template alignment algorithm that is obtained, cognitive information object, here we can think of to obtain the cognitive object of knowledge, so in this level of integration can be called knowledge fusion. In most pattern recognition systems, knowledge is expressed in the form of scores. Therefore, knowledge level fusion can also be called hierarchical fusion in pattern recognition system. Because the fusion of scores can balance the information quantity and the difficulty of handling these information in different modes, it is favored by most research institutes. In the integration of knowledge layer, the greatest difficulty comes from the heterogeneity of each score. Heterogeneity refers to the modalities from non-homologous points or different matching algorithms. For example, face and iris score comparison modal can be in the 0, 1 range, however, due to the different sources, the score distribution therefore within this interval the meaning often huge difference, even if the same score of their contents and information may also be completely opposite, and different modes of output score the classification accuracy is not the same, so it brings great challenges to the fusion. And different modes and even the same mode, because the matching algorithm used in comparison is different, the interval of the scores obtained will also be very different. As showed in the NIST Face database, someone compares scores a face matching algorithm in the interval is 1, 1, and another set of face matching algorithm that is in the range of 100 points, and the scores of different modes and different matching algorithms are different. A part is obtained by matching the similarity between the template as the score, the greater the score, the two templates are more likely to belong to the same category; the other part is the template for the distance in the feature space as the score, scores of smaller, two templates was more likely to belong to the same category. For example, in MSU multimodal database, the score of fingerprint modal output is similarity score, while facial mode output is distance score, and simple fusion will produce serious consequences. At present, there are many hierarchical fusion algorithms that can solve the above problems. They can be further divided into: transform based score fusion method, classifier-based score fusion method and density-based score fusion method [40–45].

6.3.4 Decision Layer Fusion

In the biological recognition system relies on drafts, most commercial systems only provide the final recognition results. This means that when we use the commercial system of single mode to build a multimodal biometric system, the final decision we can use to only the various modes of data fusion, in this case, only in the multimodal decision level fusion [45].

The methods of decision layer integration include:

1. and/or rule: and or rule in the research work of the Daugman have been mentioned, Vijayalakshmi also made use of fingerprint and palmprint and rule fusion. As the name implies, its operation logic is exactly the same as “and”, “or”. In the identification process, when using “and” rule, only all modes are as positive samples to determine the user, the user can be authenticated, when using the “or” rule, as long as any mode decision is the positive samples, the user is considered and a consistent statement. Although or rule may obtain W good false acceptance rate and false rejection rate, but Daugman pointed out that if the rotating Noyes equal error rate to measure the performance of the fusion algorithm, and the rule or show the effect is not good, even clever will

drop the overall performance to have led to this rule is rarely in the multimodal biometric system actual.

2. majority voting method: in majority voting method, only when more than half of the modes authenticate the user's identity, the multi-modal decision system will authenticate the user, otherwise, it will refuse the user. Like the polling voting method in the ranking layer, the advantage of majority voting is that the system does not need to acquire a priori knowledge of each modal function. It only needs to treat them equally, and there is no need for additional training in the process of fusion. The performance and performance of this method have been analyzed and discussed in detail by Lam and Kuncheva.
3. the weighted majority voting method: this method is based on majority voting method, with a weight for each mode increase. The weight of each mode represents the corresponding modal properties. The greater the weight, the more reliable the model is, and it often plays a decisive role in the voting. And the weight coefficient can be obtained by different learning methods.
4. Bias's decision fusion method: Bias's decision fusion method is developed according to Bias's law. However, similar to density based fraction fusion, the key part is not the use of rules, but the transformation of discrete class attribution to continuous probability value. Therefore, accurate estimation of probability value is the key to the success of the fusion method. The probability of Bias decision method assumes that users belonging to different classes is the same, then estimate the conditional probability of different categories, can be calculated according to the Bias rule, the user belongs to different categories of posterior probability, and then the user to the posterior probability of the largest category. Although we need to assume that every classifier is completely independent when applying Bayes rule, Domingos finds that even if this assumption is not valid, the clutch rule is still very robust. Behavior knowledge space method: the main idea of behavior knowledge space is to create a lookup table, and each item in the lookup table is a point in the decision space constituted by all modalities. When a new user application passes.
5. sorting and merging: after obtaining the contrast scores of different biological modalities, we divide the scores from the confidence level to the low order, and the resources that we merge in the ranking layer are the ordinal numbers obtained by sorting. The main work in this area is from Ho, which introduces the H method, the highest ranking method, the polling method, the logic regression method.

(1) the highest ranking in the highest ranking method, every user uses various modes sort in the highest number (i.e. ordinal value minimum) as the final number of users, and the user ranking, ranking the highest (number of minimum) users have been identified as the correct user.

(2) doda voting: the main idea of the fusion is to add all the ordinal numbers of all modalities and then sort them from low to high, and determine the correct users (the smallest number of users) according to the final sorting. The characteristic of this method is that each biological mode is assigned to the same weight, that is, the performance of each mode is considered to be exactly the same. The

advantage is that it does not need to explore the prior knowledge of various biological modalities, and does not require a learning process. Marasco used this method in its fusion work.

(3) the logical regression method: the logical regression method is improved on the basis of the polling method, and the different weights are allocated for each mode. However, the weight coefficient needs to be trained by logistic regression training. The ordinal number of the sorting layer is completely based on the contrast fraction of the corresponding modes, and the method is very simple and easy to operate. But at the same time, the ranking layer loses a lot of information on the basis of fractional level, and the final accuracy rate has different modes. The final decision must constitute one point in the space, and then identify the user through the lookup table corresponding to that point. The advantage of this method is that it takes full account of the correlation between different modes and the relative performance between each other. However, this method needs a large number of training samples as a support to build the correct lookup table, so this method is not very applicable under the condition of many users.

Because the multi-modal decision information is the easiest to obtain and intercommunicate, the research on the multi-modal fusion method of the decision layer is more extensive. However, the amount of information in the decision-making layer is often too low, which will affect the recognition effect after the final fusion [46–49].

6.4 Applications in a Robot System

The robot manages a high-level behavior of a robot through a behavior tree system, and for the key machine vision system, the advanced deep learning technology is used to achieve it. Finally, based on ROS (Robot Operating System) platform, the tight integration of multiple modules is realized, and a hybrid intelligent architecture combining autonomous mobile and high-level intelligence is built, as shown in Fig. 6.1.

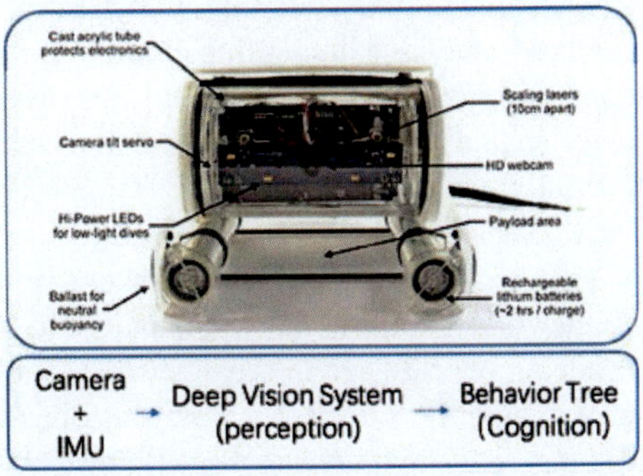

Fig. 6.1 Fusion of the robot system

ROS is a robot software platform that provides a similar operating system for heterogeneous computer clusters. ROS provides some standard operating system services, such as hardware abstraction, underlying device control, common functional implementations, inter process messages, and packet management. ROS is based on a graph structure, so that different nodes' processes can accept, publish and aggregate various kinds of information, such as sensing, control, status, planning and so on. At present, ROS is mainly supported by Ubuntu. ROS can be divided into two layers, the lower level is the operation system level described above, and the high-level is the various software packages that the majority of users contribute to achieve different functions, such as location mapping, action planning, perception, simulation and so on. The operation framework ROS is a loosely coupled architecture processing module P2P using ROS communication module of the network connection, it performs several types of communication, including synchronization based on RPC service (remote procedure call) communication, asynchronous data flow based on Topic communication, and the parameters on the server data storage.

A system using ROS includes a series of processes, which exist in many different hosts and are linked by end-to-end topology during operation. Though those software frameworks based on the center server can also realize the advantages of multi process and multi host, in these frameworks, when the computers connect through different networks, the central data server will have problems. ROS's point to point design and service and node manager mechanisms can disperse real-time computing pressure brought by computer vision and speech recognition functions, and be able to meet the challenges of multiple robots.

6.4.1 Deep Vision System

Today, a well-known construction method of artificial visual system is based on deep learning framework. First, let's look at a simple neural network, demonstrated in Fig. 6.2.

Deep learning is a new method of training multilayer neural network. Since the publication of related researches on deep network training in 2006, it has caused a boom in academia and industry, and has also produced a series of computational frameworks for deep learning. Deep learning is highly dependent on hardware environment, and has a higher threshold for developers. The appearance of deep learning computing framework shields a lot of development cost of hardware environment level, so that researchers and developers can focus on algorithm implementation and fast iteration. Currently popular frameworks include Torch, Theano, Caffe, TensorFlow, MXNet, CNTK, PaddlePaddle, etc. it is not easy to understand and master a deep learning computing framework. It usually takes weeks to months. Of these frameworks, Caffe is most convenient to use. At present, most of the paper in the academic world have the realization of Caffe, which is a good choice for the rapid prototyping. If you need to make customized development, TensorFlow and MXNet are the better two options, and Baidu's own PaddlePaddle can also be considered.

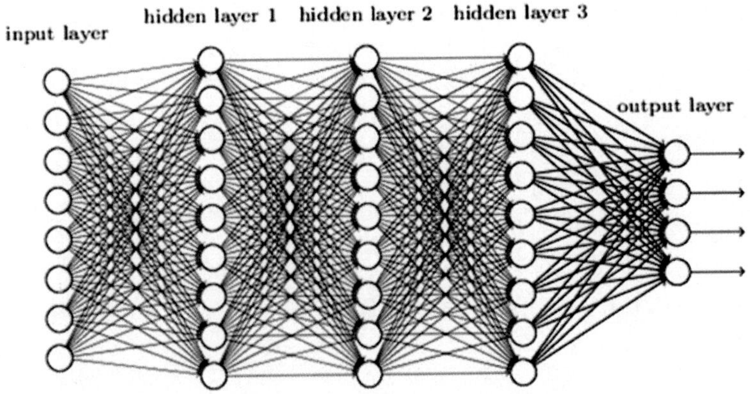

Fig. 6.2 Description of driver blind area

The neural network is composed of layers of neuron nodes. The training data is input from the input layer (input layer) node, the middle hidden layer (hidden layer) nodes do some operations, and the output layer (output layer) is connected to annotation. Then we can update the network weight by gradient descent algorithm to complete a machine learning process. In this process, there are several elements, data flow in the network, node operator responsible for operation, hierarchical network structure, and last optimization algorithm. By abstracting and encapsulating these 4 parts, Caffe forms a framework for computing deep learning. In a word, Caffe = data flow + node operator + network structure + optimization algorithm.

The deep vision system mainly completes video and image acquisition and online analysis and processing of video sensors, analyzes objects in the scope of vision, detects the objects of attention and identifies them, so as to control the movement and grasping of robots. Therefore, detection and recognition is the core function of the deep vision system. At present, there are many deep learning frameworks for detection and recognition, such as R-CNN, Fast-RCNN, Faster-RCNN, R-FCN, Yolo, SSD and so on. Among them, the SSD method is widely used, it is a combination of faster-rcnn and Yolo, combined with the regression method in Yolo, while a combination of anchorbox mechanism in faster-rcnn, SSD will output a series of discrete (discretization) bounding boxes, the bounding boxes is in different levels (layers) generated on the featuremaps and, there are different aspectratio. In the prediction phase, the object in every default box is calculated, which belongs to the possibility of each category, that is, the score, and the score. At the same time, the shape of these bounding boxes should be fined to make it fit the outer rectangle of the object. Also, in order to deal with different sizes of the same objects, SSD combines predictions with different resolutions of feature maps, and SSD method completely eliminates proposals generation, pixel resampling or feature resampling. This makes it easier for SSD to optimize training and to integrate the detection model into the system more easily.

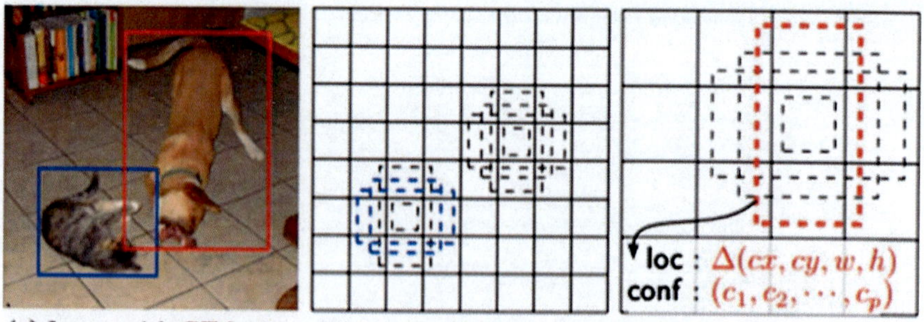

Fig. 6.3 The SSD framework

SSD also has its shortcomings. The detection rate of small targets is very low. This is because SSD preprocessed the input images. After a deep convolution layer, small targets may be lost for small targets (Fig. 6.3).

For the depth of the visual system of underwater target detection and recognition, is facing a very complex underwater environment, the illumination and atomization is serious, prone to interference and other issues, so in order to improve the stability of the system, the need for preprocessing of video data, including color correction, fog and normalization. Figure XX is an underwater target recognition effect map obtained by this system.

6.4.2 Underwater Robot System

The robot system this chapter introduces the underwater is customized based on open source OpenROV robot system and provides a basic underwater robot hardware platform OpenROV, including three engines, a high-definition camera, and some sensors, also can be conveniently installed lights and other key device. More importantly, through the ROS framework and convenient expansion of other functions, such as adding the control module more smoothly, and add the behavior more intelligent underwater etc.

The procedure of objects localization and motion planning of robots is: firstly, the object to be grasped is tracked and localized by combining the self-localization system with the vision system fixed on the robot platform, and then a valid path can be generated which guides the robot from the current position to the grasping object according to the current environment model. Taking the underwater robot as an example, the brain-inspired sites localization and object tracking are as Fig. 6.4.

Fig. 6.4 Objects localization and motion planning of an underwater robot

6.4.3 Integration Cognition with Behavior Trees

There is a new ROS package called pi_trees, which is the only behavior trees can be used in our underwater robot system.The behavior tree is a decision tree to help us define the behavior of a robot object. Conceptually, the behavior tree is relatively simple, but some of its features, such as visual decision logic, the control node can reuse, logic and implementation of low coupling, compared with the traditional state machine, can greatly help us quickly and convenient decision-making organization. The main structure of a behavior tree is its node, including two categories of behavior nodes and decision nodes. Behavior tree behavior is at the leaf node, commonly known as the behavior node (Action Node), such as the lower red circle.

These leaf nodes are the result of decision making through behavior tree, which is equivalent to a well-defined "request" (Request), such as mobile (Move), Idle, Shoot. Is the situation related behavior of nodes, due to different situations, we need to define the behavior of different nodes, but for a situation, in the behavior of the tree node is the behavior that can be reused, such as mobile, patrolling branch, need to use, in the escape branch, will be used, in this case, we can reuse this node, as shown in Fig. 6.5.

Behavior nodes are generally divided into two operating states:

(1) operation (Executing): the behavior is still in the process of processing
(2) completion (Completed): the action is completed, successful, or failed.

In addition to behavior nodes, the rest are commonly referred to as Control Node, which is represented by the scientific name of trees, such as the green nodes. The control node is actually the essence of the behavior tree. A complete behavior is defined by these control nodes (Fig. 6.6).

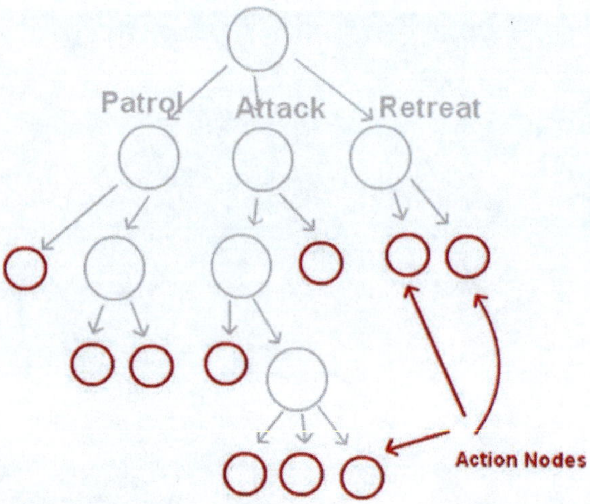

Fig. 6.5 The behavior of different nodes

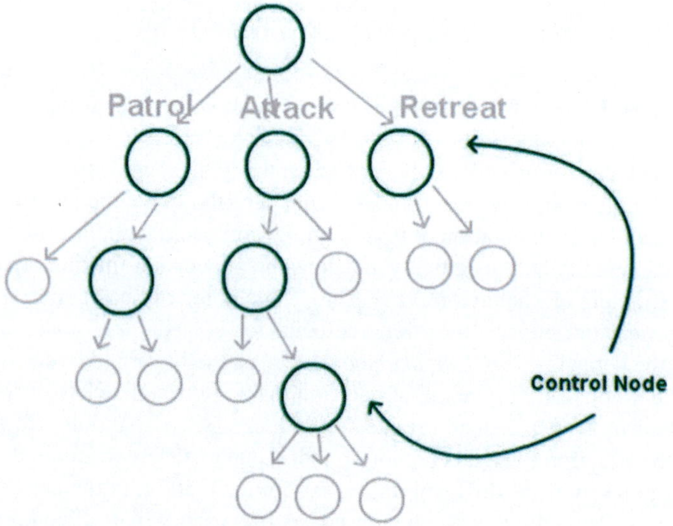

Fig. 6.6 Role of the control node

From the control node, we can see the logical direction of the whole behavior tree, so one of the characteristics of the behavior tree is its logical visibility.

Generally speaking, the commonly used control nodes have the following three kinds.

(1) selection (Selector): select a certain execution of its subnode
(2) sequence (Sequence): all of its sub nodes are executed in turn, that is, after the current one returns the "complete" state, then the first sub node is run.

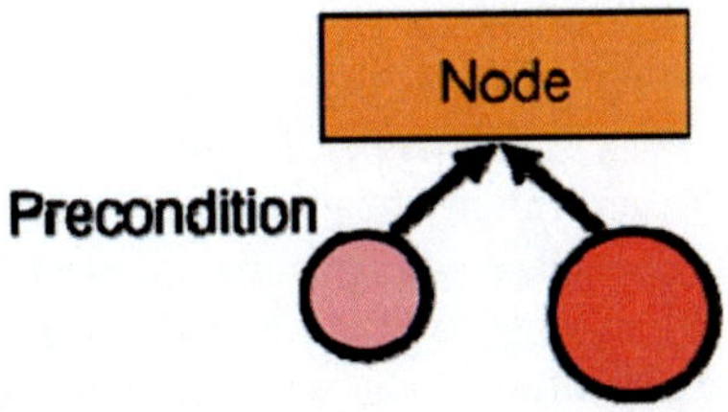

Fig. 6.7 Dependence on precondition

(3) parallel (Parallel): run all of its sub nodes.

It can be seen that the control node is actually "controlling" its sub nodes. The sub nodes can be leaf nodes or control nodes. The so-called "execution control node" is how to execute the control logic defined by it. Therefore, many other control nodes can be expanded, such as loops (Loop) and so on. Unlike behavior nodes, control nodes are independent of environment, because they are only responsible for the control of behavior tree logic, without involving any execution process.

Behavior tree also includes another concept, which is generally called the premise (Precondition). Every node, whether it is a behavior node or a control node, contains a prerequisite part (Fig. 6.7).

The premise provides the basis for "selection". It contains the conditions of entering or selecting the node. When we use the selection node, it is the premise to test every sub node in turn, and if it satisfies, we choose this node. Since we eventually return to a behavior node (leaf node), the general premise of current behavior can be regarded as the premise of the current behavior node, the premise of And, the premise of the parent node, the premise And of the parent node of the And parent. The premise of the.And root node (generally not set, directly back to True).

The behavior tree is described by the decision logic of the whole AI through the behavior nodes, the control nodes, and the premises on each node.

6.5 Application of Computer in Mixed Reality Technology

Interdisciplinary evolution of the machine brain will partly depend on the interdisciplinary application of computer in mixed reality technology. Although mixed reality has broad prospects and huge market, it can not completely replace the traditional form of entity. The application of mixed reality technology is not to deny the traditional way of physical display, but to better play its function, improve the speed of information transmission, widen the width of information transmission, and enhance the accuracy and precision of information transmission. Both of them should show their respective advantages and complement each other in communication function and effect, and jointly realize the benign role of computer information transmission and communication. Focusing on the development trend of this field in recent years, this paper focuses on the human-computer interaction design and key technology characteristics in hybrid reality, analyses the application status of different interactive technologies in hybrid reality, and finally makes a summary.Hybrid reality technology is the further development of virtual reality technology, which can enhance the authenticity of user experience by presenting virtual scene information in real scene and setting up an interactive feedback information loop among the real world, virtual world and users [50–53]. Hybrid reality technology is a set of technology combinations, which can not only provide new viewing methods, but also provide new input methods, and all methods are combined to promote technological innovation [54, 55]. The combination of input and output is a key differentiated advantage for SMEs [55–58]. In this way, hybrid reality technology can directly affect the workflow of employees and help them improve their work efficiency and innovation ability [59–61]. Hybrid reality technology will be more and more widely used in our life, changing the way people work, communicate, learn and entertain [62, 63]. The way people consume media will also enhance their participation in the world and promote communication between individuals and the world. This is in itself a change in the traditional way of life, which will make our life more exciting.

6.5.1 Development Trend of Hybrid Reality Technology

Hybrid reality technology, called Mediated Reality, is a concept of mediating reality proposed by scholars. Hybrid reality technology is the extension of augmented reality technology and the further development of virtual reality technology. It generates virtual objects that do not exist in real environment through computer graphics technology and visualization technology, and superimposes virtual objects into real environment through sensing technology. Real environment and virtual objects are displayed in the same screen or space in real time. Users use display devices. You can see a new environment with real sensory effects.

Advantages of hybrid reality technology are summarized as follows. The realization of hybrid reality needs to be in an environment where things in the real world can interact with each other. If everything is virtual, it is the field of virtual reality. If the virtual information can only be simply superimposed on real things, it is augmented reality. The key point of hybrid reality technology is to interact with the real world and obtain timely information. Hybrid reality systems usually have three main technical characteristics. It combines virtual and reality; virtual three-dimensional; real-time operation. Hybrid reality technology combines the advantages of virtual reality technology and augmented reality technology, and can better embody augmented reality technology.

In early development of hybrid reality technology, some scholars call "environment is the creative practice of information concept" as "early people used various foods in their surroundings as communication tools to satisfy their instinctive desire to express or elaborate certain experiences", and think that this is the beginning of human display design. The development of exhibition design is directly influenced by the social form and the progress of science and technology. With the development of industrial revolution, great changes have taken place in the mode of production and social formation of human society, and modern display design in a real sense has developed rapidly in this context. The power of scientific and technological progress is constantly catalyzing new breakthroughs in display design. At the same time, it establishes display design as the main form of mass information dissemination. Mixed reality technology not only embodies the power of the cutting edge of science and technology, but also is one of the new forms of contemporary multimedia art exhibition. Although its combination with display design is only a few years old, due to its broad development space and form, mixed reality display technology has developed rapidly in a few years. Nowadays, hybrid reality display technology has been applied in museums, science and technology museums, art galleries, large-scale exhibitions, TV entertainment and many computer-related meetings at home and abroad.

Development Trend of hybrid reality technology in the future is therefore conjectured. With the popularization of mobile devices, smart glasses and even helmet begin to enter the market. The carrier of hybrid reality technology has more and more selectivity. Global VR-related devices will break through 100 million, and high-end VR devices released globally include Oeulus Rift and HTC Vive. The research and development of VR equipment will inevitably promote the transition of VR to AR and MR technology more quickly. Perhaps someday in the future, everyone can see a new world in different perspectives with the help of hybrid reality technology. Some scholars believe that the arrival of mixed reality will be faster than imagination. He said: "Now the first sign of landing in the market is just the beginning. In two years, mixed reality will become a part of people's lives. People can interact directly with the images superimposed on reality in open space. In five years the world will be half a hybrid reality and half an atom. In 10 years'time, mixed reality will be everywhere.

6.5.2 Characteristics of Key Technologies in Hybrid Reality

In the application of mixed reality in display design, it is usually necessary to establish a more intuitive and natural information environment than the existing computer system. Through various kinds of sensor devices and multi-dimensional information output, it is not limited by time and space. It is based on the real world and transcends the real world. This environment has humanized input and output, and people can immerse in it, get in and out freely. It can interact naturally with it. Therefore, hybrid reality display technology has the characteristics of authenticity, real-time interaction and imagination. Compared with the previous virtual reality display technology, the greatest advantage of hybrid reality display technology lies in its simulation, reality and presence. Virtual reality can only show the virtual world simulated by computer, but hybrid reality display technology can perfect the combination of virtual world and real world, make the audience more immersed and stimulate more real sensory experience.

On the basis of the original virtual reality display technology and traditional physical display technology, hybrid reality display technology combines them as a display bridge between computer virtual and real perception world, enabling consumers to freely shuttle between virtual and real, breaking the rigidity of traditional static display and one-way information transmission, and carrying out all kinds of real-time display. Interaction and communication between people and objects. Mixed reality display technology can not only reproduce the real environment, but also arbitrarily conceive the environment that does not exist objectively or even impossible to occur. With the development of science and technology, multimedia technology and computer technology, the variety of hybrid reality display technology will continue to expand. New hybrid reality display technology can also be created by superimposing existing augmented reality (augmented reality) and virtual reality technologies. The endless development of science and technology also creates infinite imaginable space for the development of hybrid reality display technology.

6.5.3 Application of Computer in Hybrid Reality Technology

There is a huge gap between human and computer-generated virtual scenes. In the mixed reality environment, we try to construct the direct relationship between human perception and computer through the integration of virtual and reality. Therefore, the traditional interaction between human and computer is difficult to apply, and we need to adopt the natural human-computer interaction. Natural human-computer interaction is the core technology of virtual reality. Through the interpretation of human natural language and body semantics, the changes of virtual scenery can be calculated by computer and the appropriate response can be made. There are two aspects involved, one is to perceive the real behavior, and the other is to respond according to the perceived results. Interaction is reflected not only in the interaction with scenery or information, but also in the mutual understanding and feedback between users or user groups and virtual scenes. When the user interacts with the virtual crowd, the user's intention perception and the response of the virtual crowd have stronger social attributes. Natural human-computer interaction technology is the channel of reality and virtual. Perception of human interaction intention is a very important part of human-computer interaction. In the early stage, human-computer interaction was realized by various contact interactive tools such as virtual gloves, such as virtual reality gloves. Video-based recognition of human body language is also a very important development direction, without touching users, but because it is very sensitive to background motion, it is easy to be disturbed. Because of the popularity of deep video acquisition equipment, the accuracy and robustness of behavior recognition are greatly improved. The motion recognition based on the depth video skeleton is becoming more and more accurate, thus becoming a new tool for motion recognition and human-computer interaction.

The inadequacy of scene perception is an important obstacle to the integration of virtual reality. At present, the important achievements of hybrid reality technology are mainly reflected in the positioning accuracy and speed of HMD. Man perceives the three-dimensional environment, illumination changes, object movement, behavior characteristics of the real world through the five senses, and then drives the virtual objects to make appropriate adjustments according to the dynamic changes of the environment. In the mixed reality environment, whether in the real world or in the virtual world, it is also necessary to construct a reasonable scenario in order to make strong real technology have a real impact on the real world. If we can't understand the user's interaction intention, we can't make the virtual role respond appropriately. The depth of understanding human behavior determines the depth of virtual role behavior feedback. For example, positioning your hands may allow you to understand the target of boxing and generate feedback; however, it is very difficult to understand the subtle changes in human emotions and respond. Computer perception of the real world is achieved through various sensors, and the camera or depth camera is the main body of the current vision sensor. Like human vision, visual sensors can capture highly clear details. However, due to the progress of computer vision technology, the information from visual sensors is still limited. In recent years, due to the application of in-depth learning in the field of computer vision, visual technology has been able to obtain more and more high-precision perceptual information, and some abilities have even surpassed human recognition ability. Looking forward to the development of technology, visual technology will greatly promote the ability of hybrid reality technology in the field of application.

In the mixed reality environment, users may see the identification of machine parts, or a virtual cup falls on the real ground and smashes, which is the integration of virtual and real visual space, or they may see the friendly greetings of the cognitive virtual characters when they meet, which is the integration of virtual and real social space. However, the world of computers can be much more complex than the presentation of simulated scenes. The information collected in the real world is superimposed on the real world through the visualization of information after the analysis of large data. By introducing the sense of space in the real world and adding the sense of presence to the data, the understanding of the meaning of large data can be better realized. Because the hybrid reality environment has interactive function, it can provide the user-driven data function. Through the analysis and judgment of information by human intelligence, it can promote the analysis and semantic extraction of data by computer according to the user's needs, and further present the analysis results. This closed-loop analysis is the interaction of human intelligence and machine intelligence. The virtual world in hybrid reality environment is a world created by computers. Through natural human-computer interaction technology, the virtual world is provided to human beings for manipulation, and this manipulation is carried out in a manner consistent with the characteristics of human perception, thus extending human perception to the computer world naturally. From this point of view, hybrid reality can provide a platform for the integration of human intelligence and machine intelligence.

This section was explicitly reorganized from [64]. Mixed reality technology is emerging, and in the big trend, it will become the leader in virtual reality technology in the future. However, before the comprehensive practical application of the mixed reality technology, there are still many technical difficulties to overcome. The mixed reality technology from screen video to wearable helmet makes it easy for users to move and observe mixed scenes freely, thus gaining immersion. How to synthesize a mixed scene for the vision-minds brain will be a very challenging task.

References

1. Iwaniuk A N, Nelson J E, Pellis S M, et al. Do big-brained animals play more? Comparative analyses of play and relative brain size in mammals. Journal of Comparative Psychology, 2001, 115(1): 29–41.
2. Milner A D, Goodale M A. The Visual Brain in Action. 2006.
3. Sabel B A, Henrich-Noack P, Fedorov A, et al. Vision restoration after brain and retina damage: the "residual vision activation theory". Progress in Brain Research, 2011, 192(8):199–262.
4. Shortess G K. Brain, vision, memory : Tales in the history of neuroscience. Brain, Vision, Memory:tales in the history of neuroscience. 1998:150–151.
5. Arbib M A, Hanson A R. Vision, Brain, and Cooperative Computation. Mit Press, 1987.
6. Salmi J, Rinne T, Degerman A, et al. Orienting and maintenance of spatial attention in audition and vision: an event-related brain potential study. European Journal of Neuroscience, 2007, 25(12):3725–3733.
7. Lesica N A, Stanley G B. An LGN Inspired Detect/Transmit Framework for High Fidelity Relay of Visual Information with Limited Bandwidth[C]// Brain, Vision, & Artificial Intelligence, First International Symposium, Bvai, Naples, Italy, October. 2005.
8. Oksenberg A, Shaffery J P, Marks G A, et al. Rapid eye movement sleep deprivation in kittens amplifies LGN cell-size disparity induced by monocular deprivation. Brain Res Dev Brain Res, 1996, 97(1):51–61.
9. Kozak W M, Sanderson A C. Transient persistence of neural activity after periodic stimulation in the cat LGN. Biological Cybernetics, 1979, 35(4):189–195.
10. Cowey A, Stoerig P. Projection patterns of surviving neurons in the dorsal lateral geniculate nucleus following discrete lesions of striate cortex: implications for residual vision. Experimental Brain Research, 1989, 75(3):631–638.
11. Casagrande V A, Condo G J. Is binocular competition essential for layer formation in the lateral geniculate nucleus?. Brain Behavior & Evolution, 1988, 31(4):198–208.
12. Witmer L M, Ridgely R C. New Insights Into the Brain, Braincase, and Ear Region of Tyrannosaurs (Dinosauria, Theropoda), with Implications for Sensory Organization and Behavior. Anatomical Record-advances in Integrative Anatomy & Evolutionary Biology, 2010, 292(9):1266–1296.
13. Counter S A. Preservation of brainstem neurophysiological function in hydranencephaly. Journal of the Neurological Sciences, 2007, 263(1):198–207.
14. Clavio M, Nobili F, Balleari E, et al. Quality of life and brain function following high-dose recombinant human erythropoietin in low-risk myelodysplastic syndromes: a preliminary report. European Journal of Haematology, 2015, 72(2):113–120.
15. Shulman A, Goldstein B. Brain and inner-ear fluid homeostasis, cochleovestibular-type tinnitus, and secondary endolymphatic hydrops. Int Tinnitus J, 2006, 12(1):75–81.
16. Ahonniska J, Cantell M, Tolvanen A, et al. Speech perception and brain laterality: the effect of ear advantage on auditory event-related potentials. Brain & Language, 1993, 45(2):127–146.

17. Repp B H. Stimulus dominance and ear dominance in the perception of dichotic voicing contrasts. Brain & Language, 1978, 5(3):310–330.
18. Golestani N, Price C J, Scott S K. Born with an Ear for Dialects? Structural Plasticity in the Expert Phonetician Brain. Journal of Neuroscience the Official Journal of the Society for Neuroscience, 2011, 31(11):4213–20.
19. Administrator, Br W N C. Brain dynamics for perception of tactile allodynia (touch-induced pain) in postherpetic neuralgia. Pain, 2008, 138(3):641–656.
20. Cao H, Xu X, Zhao Y, et al. Altered Brain Activation and Connectivity in Early Parkinson Disease Tactile Perception. American Journal of Neuroradiology, 2011, 32(10):1969–1974.
21. Tomasello M, Rakoczy A H. What Makes Human Cognition Unique? From Individual to Shared to Collective Intentionality. Mind & Language, 2010, 18(2):121–147.
22. Schmidt C, Collette F, Cajochen C, et al. A time to think: circadian rhythms in human cognition. Cognitive Neuropsychology, 2007, 24(7):755–789.
23. Koechlin E, Basso G, Pietrini P, et al. The role of the anterior prefrontal cortex in human cognition. Nature, 1999, 399(6732):148–151.
24. Hermer-Vazquez L, Spelke E S, Katsnelson A S. Sources of Flexibility in Human Cognition: Dual-Task Studies of Space and Language ☆. Cogn Psychol, 1999, 39(1):3–36.
25. Grothmann T, Patt A. Adaptive capacity and human cognition: The process of individual adaptation to climate change. Global Environmental Change, 2005, 15(3):199–213.
26. Lewicki P, Hill T. On the status of nonconscious processes in human cognition: comment on Reber. Journal of Experimental Psychology General, 1989, 118(3):239.
27. Phillips S, Wilson W H. Categorial Compositionality II: Universal Constructions and a General Theory of (Quasi-)Systematicity in Human Cognition. Plos Computational Biology, 2011, 7(8):e1002102.
28. Stork S, Schubö A. Human cognition in manual assembly: Theories and applications. Advanced Engineering Informatics, 2010, 24(3):320–328.
29. Jr W R, Srull T K. Human cognition in its social context. Psychological Review, 1986, 93(3):322–59.
30. Ohnuki-Tierney E. phases in human perception/conception/ symbolization processes: cognitive anthropology and symbolic classification . American Ethnologist, 1981, 8(3):451–467.
31. Montague P R, King-Casas B, Cohen J D. Imaging valuation models in human choice. Annual Review of Neuroscience, 2006, 29(29):417–448.
32. Mcclelland J L, Plaut D C. Computational approaches to cognition: top-down approaches. Current Opinion in Neurobiology, 1993, 3(2):209–216.
33. Madl T, Baars B J, Franklin S. The timing of the cognitive cycle. Plos One, 2011, 6(4):e14803.
34. Guarini A, Sansavini A, Fabbri M, et al. Basic numerical processes in very preterm children: a critical transition from preschool to school age. Early Human Development, 2014, 90(3):103–111.
35. Petrini K, Remark A, Smith L, et al. When vision is not an option: children's integration of auditory and haptic information is suboptimal. Developmental Science, 2014, 17(3):376–387.
36. Evans J R. Auditory and Auditory-Visual Integration Skills as They Relate to Reading. Reading Teacher, 1969, 22(7):625–629.
37. Battaglia P W, Jacobs R A, Aslin R N. Bayesian integration of visual and auditory signals for spatial localization. Journal of the Optical Society of America A Optics Image Science & Vision, 2003, 20(7):1391–1397.
38. Arrighi R, Alais D, Burr D. Neural latencies do not explain the auditory and audio-visual flash-lag effect. Vision Research, 2005, 45(23):2917–2925.
39. Ratha N K, Chen S, Jain A K. Adaptive flow orientation-based feature extraction in fingerprint images ☆. Pattern Recognition, 1995, 28(11):1657–1672.

40. Ghadimi S, Mohtasebi M, Abrishami M H, et al. A Neonatal Bimodal MR-CT Head Template. Plos One, 2017, 12(1):e0166112.
41. Leys C, Ley C, Klein O, et al. Detecting outliers: Do not use standard deviation around the mean, use absolute deviation around the median. Journal of Experimental Social Psychology, 2013, 49(4):764–766.
42. Han C. Personal Authentication Using the Fusion of Multiple Palm-Print Features. Computer-Aided Intelligent Recognition Techniques and Applications. 2005.
43. Lin C L, Wang S H, Cheng H Y, et al. Bimodal Biometric Verification Using the Fusion of Palmprint and Infrared Palm-Dorsum Vein Images. Sensors, 2015, 15(12):31339–31361.
44. Gopal, Srivastava S, Bhardwaj S, et al. Fusion of palm-phalanges print with palmprint and dorsal hand vein. Applied Soft Computing, 2016, 47:12–20.
45. Vlasak M, Blomqvist S, Hovi T, et al. Sequence and Structure of Human Rhinoviruses Reveal the Basis of Receptor Discrimination. Journal of Virology, 2003, 77(12):6923–6930.
46. Saini N, Sinha A. Face and palmprint multimodal biometric systems using Gabor–Wigner transform as feature extraction. Pattern Analysis & Applications, 2015, 18(4):921–932.
47. Sharma P, Kaur M. Multimodal Classification using Feature Level Fusion and SVM. International Journal of Computer Applications, 2014, 76(4):26–32.
48. Ahmad M I, Woo W L, Dlay S S. Multimodal biometric fusion at feature level: Face and palmprint. International Symposium on Communication Systems Networks & Digital Signal Processing. 2010.
49. Lambercy O, Dovat L, Yun H, et al. Effects of a robot-assisted training of grasp and pronation/supination in chronic stroke: a pilot study. Journal of NeuroEngineering and Rehabilitation, 8, 1(2011-11-16), 2011, 8(1):63.
50. Chen B, Qin X. Composition of virtual-real worlds and intelligence integration of human-computer in mixed reality. Scientia Sinica, 2016(12).
51. Hamacher A, Kim S J, Cho S T, et al. Application of Virtual, Augmented, and Mixed Reality to Urology. International Neurourology Journal, 2016, 20(3):172–181.
52. Jin H, Han D, Chen Y, et al. A Survey on Human-Computer Interaction in Mixed Reality. Scientia Sinica, 2016.
53. P. Mesároš, D. Mačková, M. Spišáková, et al. M-learning tool for modeling the building site parameters in mixed reality environment. International Conference on Emerging Elearning Technologies & Applications. 2017.
54. Francia D, Liverani A. Mobile tracking system and optical tracking integration for mobile mixed reality. International Journal of Computer Applications in Technology, 2016, 53(1):13–22.
55. Kronander J, Banterle F, Gardner A, et al. Photorealistic rendering of mixed reality scenes. Computer Graphics Forum, 2015, 34(2):643–665.
56. Onime C, Uhomoibhi J, Hui W. Mixed Reality Cubicles and Cave Automatic Virtual Environment. International Conference on Ubiquitous Computing & Communications & International Symposium on Cyberspace & Security. 2017.
57. Long C, Day T W, Wen T, et al. Recent Developments and Future Challenges in Medical Mixed Reality. IEEE International Symposium on Mixed & Augmented Reality. 2017.[9]
58. Kobayashi L, Zhang X C, Collins S A, et al. Exploratory Application of Augmented Reality/Mixed Reality Devices for Acute Care Procedure Training. Western Journal of Emergency Medicine, 2018, 19(1):158–164.
59. Liu J M, Shi M T, Zhuang Y L, et al. Application of Mixed Reality in Power Grid Emergency Repair. Electric Power Information & Communication Technology, 2017.
60. Lee K F, Chen Y L, Hsieh H C, et al. Application of intuitive mixed reality interactive system to museum guide activity. IEEE International Conference on Consumer Electronics-taiwan. 2017.
61. Negri P, Omedas P, Chech L, et al. Comparing Input Sensors in an Immersive Mixed-Reality Environment for Human-Computer Symbiosis. Symbiotic Interaction. 2015.
62. Shin J, An G, Park J S, et al. Application of precise indoor position tracking to immersive virtual reality with translational movement support. Multimedia Tools & Applications, 2016, 75(20):1–20.

63. Sauer I M, Queisner M, Tang P, et al. Mixed Reality in Visceral Surgery: Development of a Suitable Workflow and Evaluation of Intraoperative Use-Cases. Annals of Surgery, 2017, 266(5):1.
64. Cong H, Ying H. Application of Computer in Mixed Reality Technology. International Conference on Frontier Computing, 2019.